ELECTRONIC STUDENT PORTFOLIOS

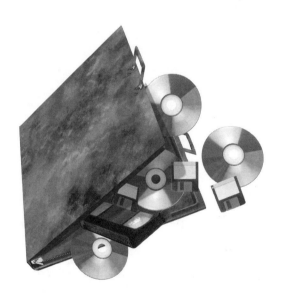

Linda E. Ash

Foreword by Kay Burke

PEARSON
SkyLight

Glenview, Illinois

Electronic Student Portfolios

Published by Pearson Professional Development
1900 E. Lake Ave., Glenview, IL 60025
Phone 800-348-4474, 847-657-7450
fax 847-486-3183
info@pearsonpd.com
http://www.pearsonpd.com

ISBN 1-57517-281-X
LCCN: 00-131513

2668-V

ZYXWVUTSRQPONMLKJIHGFEDCB
06 05 04 03   15 14 13 12 11 10 9 8 7 6 5 4 3 2

SkyLight Professional Development

TABLE OF CONTENTS

ACKNOWLEDGMENTS

I would like to thank N. Gerry House and Dale Kalkofen for their leadership in school reform. In my search for a deeper understanding of the integration of technology across all curricula and how to transfer this to children, I want to thank Deborah Gunn for her endless creativity and insightful guidance.

I want to extend a heartfelt thank you to Jim Luckey who encouraged and believed in me from the first day I appeared in his office. Many thanks and appreciation to Robin Willner, Patti Scanlon, Sarah Weber, Kevin Ho, Lisa Shepherd, Jean Ward, Diane Raley, Bill Byles, Mazie Marlin, Betty Buchignani, Celia Moore, John Avis, and Jena Pennington and her students at Oak Forest School.

Last, but not least, I thank my husband David and son Steve for their unconditional love, support, and technical advice.

I am forever indebted to all of the above.

LINDA E. ASH

Software Acknowledgments

HyperStudio is a registered trademark of Roger Wagner Publishing, Inc.

Kid Pix Studio© Broderbund Software, Inc. 1996. All Rights Reserved.

Inspiration and Inspiration Software are registered trademarks of Inspiration Software, Inc.

Neighborhood Map Machine is a registered trademark of Tom Snyder Productions, Inc.

Microsoft Works, Microsoft Word, Microsoft Access, and Microsoft PowerPoint are registered trademarks of Microsoft Corporation.

Claris Home Page and Claris Works are registered trademarks of Claris, Inc.

FOREWORD

An electronic student portfolio integrates authentic learning, assessment, and technology to provide a more accurate portrait of the student as a learner. It showcases the student's progress in meeting standards, attaining understanding of key concepts, and mastering the technological skills essential for success in both school and life.

In *Electronic Student Portfolios,* Linda Ash utilizes both written and visual steps to introduce the development of instructional or working portfolios and assessment portfolios used for evaluation. She describes the hardware and software options and the technical skills necessary for teachers and students to produce meaningful products and performances. In addition, she addresses practical issues regarding room arrangement, safety issues, scheduling, grouping, job assignments, and organization of projects and units. She also provides examples of students' work, rubrics for assessing the quality of the work, and a sample of an integrated unit.

More importantly, however, Linda shows how students can assume responsibility for collecting, selecting, and reflecting on their own work and communicating electronically with a wider audience to share their progress in meeting academic goals. Electronic portfolios provide a showcase for written work, pictures, audio and video segments to document a student's growth over time. As teachers begin to move away from traditional notebook or pizza box portfolios and experiment with CD-Roms, they will become more efficient in capturing students' authentic work from grades K-12 without cumbersome storage or organizational problems.

Educators everywhere will welcome this resource and recognize its potential to link technology to standards, instruction, and performance assessment. Linda Ash has created a valuable tool for teachers to help students communicate electronically the progress they are making in their educational journey.

KAY BURKE

June, 2000

INTRODUCTION

Today's technology and the demands of a standards-based curriculum have put a burden on many teachers. Pioneering new technologies and using multiple assessment strategies are difficult without a plan or prototype to follow. Guidance and direction reduce resistance and open the doors for change. Educators will find that they possess much of the knowledge and many of the skills needed for this type of project. Even the equipment needed to start is often already within a teacher's grasp but is going un- or under-used. This book will help educators make use of the human and material resources they have at hand to institute an electronic portfolio system in which student progress can be communicated through tangible examples.

Each of *Electronic Student Portfolios'* seven chapters contains a "Plugging in the Portfolio" section showing examples of successful electronic portfolio projects. These examples demonstrate how teachers at all levels of technical expertise can facilitate and guide students in the creation of electronic portfolios. Various formats illustrate how a traditional paper portfolio can be transformed with technology. A unit plan based on electronic portfolios is offered in an appendix as are a number of valuable Web sites that will help teachers "plug in."

Chapter 1 sets the stage for why there is an urgent need to implement electronic portfolios into the K–12 teaching and learning process. It discusses the relative benefits and purposes of different types of portfolios. The chapter also discusses ways for teachers to self-assess their levels of technological expertise. The five progressive levels defined here are touched upon in succeeding chapters as well as in the sample portfolios.

Chapter 2 explains the hardware and software needed to begin. (A glossary of technical terms at the back of the book will help navigate the chapter's terminology, as will the explanations throughout the chapter.) Project expansion is discussed and correlated to professional development as teachers move up through the levels of expertise. The "Working with Existing Equipment" section provides both strategies for educators to start now and encouragement to those who feel they must wait because they do not have the latest and greatest equipment.

Chapters 3 and 4 provide options for educators who have the basics under their belt and who need concrete implementation strategies for all grade levels. Different ways to organize student products makes use of the collection, selection, reflection, and projection processes. Guidelines for each process and different age groups are discussed.

Vision categories are introduced in chapter 5 to link essential skills to content-area standards. As students gain knowledge in specific content areas, they do not always use process skills effectively to demonstrate understanding. Vision categories help transfer appropriate skills across all curriculum areas.

All classroom activities need to be organized and managed if learning is to take place. Chapter 6 discusses effective project organization, room arrangement, safety issues, scheduling, grouping, and job assignments at grade-appropriate levels. Chapter 7 aligns curriculum, instruction, and assessment to measure student learning. Discussions include how electronic portfolios benefit all populations of students by providing information to guide them as they become more responsible for their learning and play an active role in communicating what they know to various audiences.

Ultimately, the purpose of this book is to inspire teachers, regardless of their level of technical expertise, to help students creatively design electronic portfolios that communicate the progress they are making in their educational journey. No doubt it is easier to collect papers in a pizza box, but think of the skills students miss. Administrators may say that they do not have time to implement an electronic portfolio or that they cannot afford the computers. But, can we afford *not* to do so?

WHERE TO BEGIN

Why Implement an Electronic Portfolio System?

A portfolio is a collection or display of artifacts that has been gathered systematically to demonstrate one's skill level, growth over time, or understanding of a particular concept or discipline. The use of portfolios has existed in many fields for a long time. Professionals use portfolios to display their accomplishments, testimonials from previous clients, and letters of reference. As new work is completed, the portfolio is updated, reorganized, and redeveloped to meet current needs. Such an approach to showcasing student achievement has been growing in the education sector for more than a decade.

Traditionally, educators have given multiple-choice tests to measure students' ability to recall factual information, depended upon letter grades to communicate student progress to parents, and utilized standardized tests to evaluate the effectiveness of curricula. Although these evaluations provide important information about student learning, they are no longer sufficient. Standardized tests cannot measure students' ability to problem solve, analyze, and apply what they know to new situations. As students complete their educational career and enter a work force in the *information age,* not only must they obtain content knowledge, processing skills, and good work habits, they must also

Plugging in the Portfolio

The sample portfolio in this chapter represents one created when teachers are at the Entry level of expertise. At this level, teachers have minimal involvement with technology and struggle to use technology in the classroom. They rarely schedule time for students to work on computers and prefer that a computer lab teacher deal with anything related to computers. The sample followed throughout this chapter uses a single word processing file, adding information in journal form. It is considered teacher-centered since it most likely is constructed in the computer lab. Also, at this level teachers generally limit portfolios to individual work, since classroom management strategies are more complex when groups of students begin working on a group or whole class portfolio.

be able to apply what they have learned to real-world situations and to use information in meaningful ways.

Many teachers and students want to create links among goals, standards, student work, and vision categories within a multimedia format (i.e., video, audio, text, image). More and more districts are assessing student performance through authentic work, the same way professionals demonstrate their abilities by compiling samples of work in a portfolio. Unleashing the power of portfolios will transform how educators measure what students learn.

The use of portfolios enables students to include a variety of work samples. Because of the opportunity portfolios afford for students to demonstrate what they know in a variety of ways, teachers are better able to diagnose each student's strengths and weaknesses instead of just comparing and ranking one student against another. Consequently, the emphasis is shifted away from outcome-based assessments, where students are given a test when they complete a unit to determine what factual information and basic process skills they have learned as a result of

> **Please note:** student work shown in the Plugging in the Portfolio samples may contain spelling errors indicative of student work in progress.

Plugging in the Portfolio

SKILL LEVEL

- ☒ **Entry**
- ☐ **Adoption**
- ☐ **Adaptation**
- ☐ **Appropriation**
- ☐ **Invention**

```
File   Edit   Move   Tools   Objects   Colors   Options   Extras   Help
```

What Do I Say?
Draft
By: Melissa Baker

I see a criket in yellow stuff. It is amber from treess The circket was probably bron in that tree many years ago. I got stuck in this tree and that is why he is there and that is why that is whu he got sutk. I think the cricket has been there for a very long time. The crickt has been there since the dinosaur era. When the cricket died he was covered in sap the cricket has turned into a fossel this fossil is diffeent fron most fossels because it is not embedded in a rock it is covered by tree sap

02-16 Melissa: Check with Mrs. Bradley on grammar, mechanics, and content. I suggest you use Spell Check to catch spelling errors. I would also suggest double spacing.

SAMPLE 1.1 **The student uses word processing to create a rough draft. The teacher then places a comment at the bottom of the page before giving the diskette back to the student.**

their study. In such assessments, the focus is on low-level recall of facts rather than the process progress the student has made over a period of time. Outcome-based assessment disregards the fact that students learn at different rates and ways and that all students learn and can be successful. This change is also driving a shift in attitudes toward the role of assessment in student learning. Evaluative feedback and reflection, when integrated with instruction, can actually enhance student learning and, consequently, performance. The best practice merges performance-based learning and assessment so that learning can be examined but not disturbed. Portfolios facilitate this.

To accomplish instructional change that results in improved student achievement, educators must examine their assessment needs, focusing on improved student achievement by helping students build relationships and knowledge. Implementing electronic portfolios is a way to achieve these goals.

Multiple assessment strategies give teachers insight and allow them to answer the question, *How well can students use what they know?* As multiple assessment strategies are more commonly practiced, educators grapple with how to communicate about these performances to parents and how to store this new kind of student work. Some educators compile examples of student products and performances and place them in student portfolios to provide a tangible account of student progress over time. Whether it is information learned from a field trip, from listening to a guest speaker, or from a unit of study, these tasks can be stored in a portfolio to capture the process and growth that has occurred.

Why then should an electronic student portfolio be implemented? Students' progress must be assessed by looking at multiple examples of their work. How artifacts are assessed and communicated to an audience tells students what is most valued. Electronic portfolios are also important instruments for communicating technology skills needed for the twenty-first century and for laying the groundwork for student-led conferences, which are vital if students and parents are to understand learning goals and see progress. In addition, digitizing the portfolio lessens or eliminates certain issues related to using portfolios in the classroom, including the cost of implementation, the administrative difficulties for the teacher, the practicality of its dynamic environment, and the subjectivity of scoring the work.

Cost

The cost of a pizza box (a widely used means for storing portfolio materials) is approximately $2.00 per student. On the other hand, implementing an electronic portfolio system is often a matter of putting the resources that have already been purchased to a productive use. Further, the program can be used year after year and gives students a tool to extent learning in a seamless way. With the billions of dollars spent on placing computers in classrooms and connecting all schools to the Internet, accountability will soon be a consequence. Too much money is being spent to let the equipment stand idle or be used for noninstructional activities. School districts and the businesses and communities that support them are going to demand true technology integration into all areas of the curriculum to justify the expense. Electronic portfolios provide not only justification for expense on technology but also evidence that real learning is taking place.

Administrative Difficulties

Electronic portfolios require teachers to transfer technology tasks to students to make them practical and successful. Creative ways to accomplish this must therefore be put into place. Peer tutoring, mentoring by older students, and extra help from community members will help accomplish the task. It may be easier to collect papers in a pizza box, but think of the skills students are missing. Schools may say they do not have time to implement an electronic portfolio or that they cannot afford the computers. But, can educators afford *not* to do so?

Practicality

Electronic portfolios are a practical approach to storing student information. Demonstrating what one can do results in products or performances that come in all shapes and sizes. Digitally storing this information on a CD, Internet server, or hard drive reduces the chances of misplacing the work, takes up less space, and is potentially more permanent.

Subjectivity of Assessment

As more districts implement a standards-based curriculum and assessment, they rely less on standardized, objective tests and more on authentic assessments that enable students to demonstrate understanding by performing a task. Portfolios are used to collect work and provide an organized approach to conferences. By looking at products and performances rather than just a number in a grade book and trying to determine what that number stands for in relation to the child's level of understanding, parents gain a better insight into their child's knowledge and skills.

Matching the Purpose to the Portfolio

The use and focus of the portfolio determines what kind best houses the artifacts that should be included. There are two categories into which portfolios fall—instruction and assessment. Defining the purpose for using portfolios in the classroom begins with teachers asking themselves, *How and why will I use portfolios with my students?* Once this question is answered, the purpose becomes clear, and it is easy then to know where to begin. Someone thinking about undertaking a portfolio system should ask him or herself the following questions:

1. Is this portfolio going to represent several disciplines or a particular classroom project?
2. Is there a standard set of artifacts each student's portfolio must contain at specific grade levels in order to show a spiraling curriculum?
3. Are the portfolios for instructional or assessment purposes?
4. Will the collection of work represent one event, one project, one year, or multiple years of study?

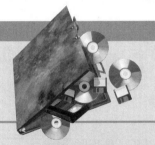

Plugging in the Portfolio

File   Edit   Move   Tools   Objects   Colors   Options   Extras   Help

What Do I Say?
By: Melissa Baker

I see a cricket covered in yellow amber. Amber matter comes from trees. Many years ago, this cricket was in a the wrong place at the wrong time and may have gotten stuck. The cricket has been there since the dinosaurs ruled the earth. This is a fossil, but a different kind. Most fossils are embedded in rock. This ricket is not in a rock but covered in tree sap.

SKILL LEVEL
☒ **Entry**
☐ **Adoption**
☐ **Adaptation**
☐ **Appropriation**
☐ **Invention**

SAMPLE 1.2 At the Entry level, students can utilize the expertise of the computer lab teacher to create a final draft of an assignment for inclusion in their portfolio.

5. Will the portfolios be used to assess student progress across the curriculum or in specific areas of study?

6. Will the portfolios be used by students or teachers to determine what courses of study to offer or what courses a student should take next?

7. Will the portfolios be used to demonstrate competency to determine graduation eligibility?

8. Are the portfolios going to showcase students' best work during a particular school year or throughout the students' K-12 education?

9. Who selects the artifacts to be included in the portfolio? What criteria are used?

10. Will artifacts collected from outside the school be included in the collection?

11. Will the portfolio represent the whole class, a small group, or an individual performance?

Answering these questions in detail will determine the purpose for integrating portfolios in the classroom. Defining the purpose for the use of portfolios helps students understand the need to think about their work and to know the standards of quality against which their work is being measured.

Types of Portfolios

The use and focus of the portfolio determines what kind would best house the artifacts that should be included. There are two categories into which portfolios fall—instruction and assessment.

Instructional Portfolios

An instructional portfolio shows assignments and tasks including draft forms, brainstorming notes, drafts, rewrites, comments from teachers and peers, and self-reflections. Assignments demonstrating certain process skills may also be placed in this portfolio to determine any further instruction that may be necessary prior to performing the culminating task. All of these tools lead the student toward products that may be placed in an assessment portfolio.

Using instructional portfolios in the classroom documents student growth over time. This type of portfolio can improve student self-worth and serve as a communication tool with parents. Sometimes referred to as a *work-in-progress* portfolio, it is organized based on standards for learning objectives. Each student work sample helps define and provide feedback on the level of performance. Most pieces of work are in draft form and have a purpose for inclusion guided by the standard. In addition, instructional portfolios help drive content instruction based on the needs of students. For example, if students are asked to write a persuasive letter, they demonstrate how well they understand the concept of persuasion. If a teacher reviews several students' letters and reflective comments, the teacher can determine whether or not they understand the concept or if it needs to be retaught. Other examples of ways instructional portfolios help to meet the needs of students include the following:

- Creating a literacy balance among reading, writing, and thinking
- Engaging students in self-reflection
- Providing students with a sense of motivation, ownership, participation, and accomplishment.
- Extending the time devoted to authentic practice writing

Advances in technology allow instruction to be personalized to reflect the goals and standards each student has demonstrated through selections placed in student portfolios. Without technology this task would be difficult to manage and too time-consuming for teachers. Technology gives teachers flexibility for where and when they view student work—especially if the portfolio is Web based. Communication will continue to become easier with the increased use of e-mail. When teachers want to discuss an assignment with parents, the electronic file can be attached to an e-mail message and sent to them. They not only can view the student work online but can also respond to the teacher's message regardless of where they are physically located.

Because many work-in-progress portfolios serve as temporary holding places for student work, it may not be an efficient use of time if it takes an extra step to transfer artifacts into a digitized form. If a word processing program is used for writing, it may benefit the author if "proofing marks" can be added in a different color and then accepted or rejected at a later date. In addition, technology is especially useful for work-in-progress portfolios, if peers outside the classroom or school will critique the work and recommend revisions. An electronic format would make access much easier. Two types of work-in-progress portfolios—journals and unit/theme portfolios—are discussed here.

Journal

Purpose: To determine if progress is made at appropriate levels

Work in a journal can be ongoing and of indefinite duration and can be used across disciplines with individuals, small groups, or with the whole class. These portfolios are ongoing logs or diaries consisting of entries such as story writings, personal reflections, self-assessments about an activity, or writings to specific prompts. A journal or writing log allows students to maintain ongoing reflective records related to a specific skill or discipline when examining their progress. Journals can also be used to expand communication between the student and teacher. Technology can make the process easier by eliminating all the paper shuffling. One example would be to create journals using e-mail and saving the messages in a folder on the server.

Unit or Theme

Purpose: To deepen students' understanding of conceptual ideas that transfer across time and cultures.

The duration of time spent using a unit or theme portfolio can vary widely (a week, a month, an entire grading period). Time spent should be long enough to sufficiently cover the subject at hand. It can be used across all content areas. An example of a theme portfolio might be Native American Populations. Individuals, small groups, or the whole class could create this type of portfolio.

Assessment Portfolios

An assessment portfolio includes work showing the level of understanding or mastery demonstrated by the student. Developmental progress is shown for a specific period of time along with self-evaluations of specific tasks. An important component in this process is the student's ability to reflect upon his or her work. Comments that illustrate self-reflection as well as those by teachers and peers will focus on the specific curriculum outcomes. Following is an explanation of the types of assessment portfolios.

Event

Purpose: To promote students' reflection upon their participation in an event, such as a field trip, and the learning that occurred as a result.

An event portfolio represents a specific occurrence or outing. It measures what was gained by the experience. Many times it is created after the experience. Technology is an excellent way to communicate the experience to parents, teachers, peers, or community members. The portfolio could also be placed on display at that location. Integrating technology into the learning process motivates students to reach their full potential. Event portfolios can be employed in a single discipline or across disciplines.

Project

Purpose: To plan, collect, and record information for all stages of the project to give students an authentic application for knowledge learned in content areas.

A project portfolio represents a deep level of understanding when the student applies knowledge and skills to real-world projects. The integration of technology is vital if students are to be prepared for the workforce of the twenty-first century. Students can use computers to extend their learning through simulations, researching on the Internet to access global information, collaborating with experts around the world, or creating multimedia presentations. Artifacts can be added to the portfolio on a daily or weekly basis for the duration of the project.

High Stakes/Graduation

Purpose: To plan, collect, and record information for all stages of the project.

"Graduation by Exhibition" is becoming more common in schools across America. It allows students to demonstrate mastery of the standards outlined for graduation. This eliminates the idea that students will receive a diploma for "seat time" in the classroom regardless of what they know when they leave school. As more and more jobs are based on one's ability to access and use information, technology skills must be demonstrated. A graduate portfolio may be one of the many requirements students must meet in order to complete their K–12 education. For example, to meet a language arts requirement, a student must successfully complete a persuasive, descriptive, and narrative essay, a graphic organizer, and a major comparison report. The electronic portfolio can be the vehicle they use to demonstrate that competency has been met. Graduation portfolios are organized so that the reader can easily view the final product in each section. If drafts are included, they should be arranged in date order with the students' reflection and scoring rubric.

K-12 Career Portfolios

Purpose: To meet criteria for graduation. Allows student to reflect upon the learning that occurred as a result of their K-12 education.

Student growth should be assessed from kindergarten through graduation. In order to continue to add information throughout a student's K–12 education, creating an electronic portfolio is important in the storage process. With the high mobility rate among students, Web-based storage is a must if the portfolio is to follow the student.

Whole Group

Purpose: To display collaborative group work. Allows for reflection upon student work as a whole class.

This assessment portfolio measures one's ability to work within a team, communicate, problem solve, and form conclusions—many skills needed to function in the workforce of the twenty-first century. Since many of these skills are observable or embedded within other work that creating a separate electronic portfolio to document these may not be worth the trouble. All students in the class would contribute work to the portfolio.

Subject Area

Purpose: At the high school level for students to demonstate knowledge of specific content knowledge.

A subject area portfolio (also called a content portfolio) is a collection of work representing the learning in a specific discipline. The integration of technology is vital for students to compete for high-paying jobs when they finish their education. It can be either instruction or assessment driven, depending upon the type of artifacts included.

Showcase/Proficiency

Purpose: To demonstrate what level of achievement a student has attained.

This portfolio displays students' best work, demonstrating to what level a standard has been achieved. Students and/or teachers choose what is included. Often this type of portfolio is used for high-stakes assessments for promotion or graduation. According to Lankes (1995), students at Central Park East Secondary School in New York City are required to complete fourteen portfolios to demonstrate their competence and performance in areas such as science, technology, ethics and social issues, community service, and history.

According to Danielson and Abrutyn (1997), educators find that a portfolio system is an essential tool needed to assess the success and merit of instructional strategies and the achievement of demanding standards because they serve to do the following:

- Engage students in a deeper understanding of the content beyond the facts.

  Students are engaged in deeper understanding of content when an emphasis is placed on interactivity throughout the learning process. The use of primary resources to obtain information has increased rapidly as Internet access continues to become more available in classrooms. As traditional textbook assignments (e.g., read the chapter and answer the questions) are less common and activities requiring students to gather their own information, solve problems, and form conclusions continue to increase, the use of portfolios will make the task manageable for both the teacher and student.

- Help students learn the skills of reflection and self-evaluation.

  Interaction among peers and teachers during teaching and learning is connected to the brain-compatible philosophy that states students create their own knowledge when engaged in active, meaningful experiences (Ronis 1999). Giving students the opportunity to reexamine their work and critique peers' work as part of the portfolio process develops a deeper understanding about the relevance and importance of the content.

- Measure learning in areas that do not lend themselves to traditional assessments.

  Traditional paper and pencil tests cannot assess many skills students need to be successful both today and into the future. Strategies such as cooperative learning, problem solving, or the ability to integrate technology into the learning process can only be assessed by viewing performance.

- Facilitate communication with parents.

  Students can use their portfolios to communicate what they have learned to parents as well as to peers (one-on-one or in groups), community members, and teachers. Transferring a traditional portfolio into an electronic one can expand communication across the city and around the globe. For example, if a student created a writing portfolio using a word processing program, that file could be e-mailed to a parent or relative living miles away.

The Need for Technology Skills

The need for persons with technology skills continues to be great. To meet this need, teachers who encourage students to use technology as a tool to process and demonstrate knowledge promote content learning *and* technological expertise. Teachers' classroom practice should promote this by doing the following:

- Maintaining a vision to achieve constant student-learning goals
- Implementing a system of curriculum, instruction, and assessment
- Focusing on teaching strategies that produce quality student work
- Developing students' self-assessment skills as part of their learning process

- Meeting district or state goals or standards
- Recognizing and encouraging the development of multiple learning styles and intelligences for *all* students
- Integrating instruction and assessment
- Encouraging student self-assessment through the process of self-selection/reflection from the products and performances created
- Enabling students to integrate technology as a tool to complete their work
- Using technology as a vehicle for communication with teachers, students, and parents
- Encouraging collaboration among class members to create a class profile
- Aiding students in demonstrating criteria necessary for graduation
- Showing student growth patterns of skills and depth of knowledge over time

The Role of the Electronic Student Portfolio in an Overall Assessment Plan

With the purpose defined and any district or state mandates taken into account, it is important to look at how student portfolios, and more specifically ones in an electronic format, will fit into an individual teacher's plan for assessing students. Will an electronic portfolio become an additional way to evaluate student performance along with traditional methods or will it be another tool used within an existing system of curriculum, instruction, and assessment? These issues depend upon the teachers' level of expertise in the areas of standards-based curriculum, project-based teaching and learning, the integration of technology, and performance assessment.

The electronic portfolio should play a major role in students' overall assessment plan, ensuring that the standard taught is also what is being assessed. In the planning stage, teachers should think about the kinds and level of technology available to their students, then reflect upon their own self-assessment professional plan, especially in the areas of technology and assessment. Electronic portfolios should not replace traditional assessment methods but instead add to them to get a truer picture of student performance. When fully implemented, they can demonstrate the *what, why,* and *how* of student learning, providing concrete representations of student experiences, performances, and products.

What Technology Skills Are Needed to Transform Student Portfolios into Electronic Ones?

Teachers and students must acquire new skills in order to transform the environment from traditional textbook, rote memory to an active, student-centered environment where tools such as technology are used to access information and use it in meaningful ways. The phrase, *show what you know* lends itself to creating electronic portfolios in all areas of teaching and learning. Electronic portfolios become the natural bridges for students to achieve goals and the standards to prepare them for success. Before teachers can transfer knowledge and skills to their students, they must first establish a baseline of their own skills in the area of technology. A five-level teacher profile was identified in the February, 1999 CEO report on School Technology and Readiness based on a study by Apple Classrooms of Tomorrow (ACOT). This study used a decade of research on instructional change as the integration of technology occurred in classroom practices.

Plugging in the Portfolio

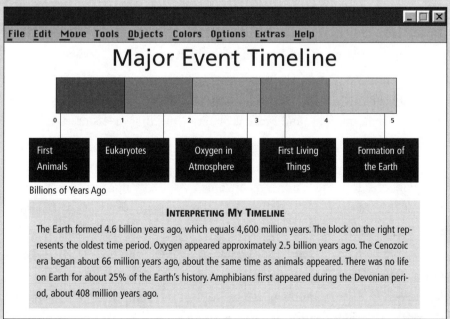

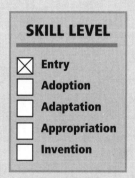

SKILL LEVEL

- ☒ Entry
- ☐ Adoption
- ☐ Adaptation
- ☐ Appropriation
- ☐ Invention

File Edit Move Tools Objects Colors Options Extras Help

Major Event Timeline

0   1   2   3   4   5

| First Animals | Eukaryotes | Oxygen in Atmosphere | First Living Things | Formation of the Earth |

Billions of Years Ago

INTERPRETING MY TIMELINE

The Earth formed 4.6 billion years ago, which equals 4,600 million years. The block on the right represents the oldest time period. Oxygen appeared approximately 2.5 billion years ago. The Cenozoic era began about 66 million years ago, about the same time as animals appeared. There was no life on Earth for about 25% of the Earth's history. Amphibians first appeared during the Devonian period, about 408 million years ago.

SAMPLE 1.3 During computer lab class, students are asked to bring data to learn how to create a chart (timeline) using the word processing program.

Entry

Teachers are not using technology themselves. If students are using technology in the classroom, they have learned it outside of the classroom
environment from someone other than the teacher. Students either
attend a computer lab class without their teacher's participation or use
educational software on a rotating basis on a classroom computer.

Adoption

Teachers use technology on a limited basis to perform tasks they would
ordinarily have done without it. They begin to see how this tool makes
traditional tasks easier, such as writing a form to send to parents for a
field trip. Teachers provide opportunities for students to use word processing to complete their assignments.

Adaptation

Teachers begin making connections between technology and curriculum.
Web sites are used to present information on a particular topic or unit of
study. Internet usage is limited to sites the teacher has discovered during
planning.

Appropriation

Teachers begin to see how the integration of technology can improve
student learning to achieve desired outcomes. Students are given more
opportunities to use problem solving and critical thinking skills through
the use of technology as a tool in their learning.

Invention

Teachers transform their classroom to take advantage of the power technology brings into the environment. Semester or yearlong projects are integrated throughout the curriculum. Students use various forms of technology to research, produce, and present quality work that meets or exceeds
the standard.

Figure 1.1 provides a means by which teachers can determine their
current level of technology skills—their comfort level, so to speak. In addition, the figure allows for easy navigation through this book. Even at the
Entry level, a teacher can begin an electronic portfolio project with students. Do not let a novice technology level block the opportunity to grow.
Spend ten to thirty minutes a day in front of the computer, navigating
the desktop until a comfort level is reached. Create lesson and activity

SELF-EVALUATION OF TECHNOLOGY SKILLS

Directions: Use the following descriptions of the five levels of expertise to determine your current level and record it in the space provided. Periodically reevaluate your level.

Invention – Congratulations! You have mastered the technology skills and are facilitating a learning environment that will enable students to successfully create all types of electronic portfolios to achieve benchmarks and goals. Collaborative learning experiences occur regularly, allowing students to develop teamwork, communication, and higher-order thinking skills. After gaining the "how-to" information from reading this book, start your project with any or all of the equipment detailed in chapter 2!

Appropriation – You have the technology skills needed to begin almost any type of electronic portfolio project. The scale of the project depends on the current amount of time students spend collaborating and working in groups to solve problems. If students currently engage in project-based learning, one strategy would be to start with a project portfolio to document a unit of study. Strategies gained from this book will help facilitate and manage ongoing group collaborations. As electronic portfolios support continuous instructional improvement, students will become more collaborative, interactive, and reflective in their learning.

Adaptation – You are discovering the potential for increased learning opportunities using technology and can begin a successful electronic portfolio project in the classroom. At this level it is recommended that students begin with a writing portfolio. As the teacher gains understanding of different software and hardware as detailed in chapter 2, it will be easy to continue to expand the project as the teacher moves to higher levels of expertise in the area of technology integration.

Adoption – Since you are just getting over the initial struggle of basic computer operation and correlating drill and practice software to classroom instruction, begin with a journal portfolio. Students can add information on a continual basis to enhance personal productivity. As access to technology increases and you gain more software application knowledge, you can move to the Adaptation level and expand the electronic portfolio project.

Entry – Give yourself a hand for taking the plunge into technology! Lots of information will be gained from this book. As basic "how-to" technology skills are acquired, begin an electronic journal portfolio to correlate with lessons and activities planned for the classroom. Here are some guiding questions to think about during planning:

- Examining the lesson just created, how could the technology enhance instruction? If the equipment is not currently available, what is needed?
- How could students use technology in this lesson to extend their learning?
- What software is available to support extra skill practice if needed by students? Where is it located?

Self-Assessment of Technology Skills
Current Level: _____ Date: _____
Evidence to Support Choice: _____

Figure 1.1

plans as usual. Once the students have performed the task, direct their attention to the large screen monitor, if one is available; otherwise, gather small groups of students around the computer on a rotating basis. Open a word processing program to begin a whole class journal portfolio. Students can take turns adding journal entries after each important activity throughout the year. As both teachers and students overcome the initial struggles, expand the project by enabling each student to create his or her own individual electronic journal portfolio.

Plugging in the Portfolio

SKILL LEVEL

☒ **Entry**
☐ **Adoption**
☐ **Adaptation**
☐ **Appropriation**
☐ **Invention**

| File   Edit   Move   Tools   Objects   Colors   Options   Extras   Help |

AutoB & Title Cover.lwp

descriptive writing sample.lwp

narritive writing.lwp

Science weather experiment.lwp

Table of Contents.lwp

SAMPLE 1.4
Using a word processing program, students view a screen with file names to determine what assignments have been placed in the electronic portfolio.

FITTING THE TOOLS TO THE TASK

Making the Move from Hard Copy to Electronic Format

As with the creation of any product, the artist, author, or craftsperson must have the necessary tools to carry out a project. However, even if a classroom is equipped with only one stand-alone computer and some word processing software, student masterpieces can be created. By beginning with an inventory of existing technology, teachers can determine what capabilities they have and how they can begin. Figure 2.1 provides a rubric teachers can use to help determine the limits or potential in their access to technology. The results of this examination should be looked at in conjunction with the individual teacher's personal level of technological integration described in chapter 1.

Professional Development

In order for an electronic portfolio project to be successful, teachers must engage in a new kind of professional development. It must be directly linked to the students' products and promote sustained training and follow-up for teachers. To be effective, professional development plans must be based upon current technology skills. Teachers can determine what stage of technology adoption and integration they fall into using the levels of technologcal inte-

Plugging in the Portfolio

The sample portfolio in this chapter represents one created when teachers are at the Adoption level of expertise. At this level teachers begin assigning students to work on computers on a regular basis. Teachers have begun to take an active role in instruction of technology related to lessons, and they are using basic applications for themselves. A teacher at the adoption stage has either attended a workshop to learn how to use PowerPoint presentation software or has asked the computer lab teacher to create a template for students to copy onto their diskettes to use as a framework to begin their electronic portfolio. This teacher-centered portfolio is a sample of how an individual student can place information into the teacher-made template. At this level teachers generally limit portfolios to individual work since classroom management strategies are more complex when groups of students begin working on a group or whole-class portfolio.

TECHNOLOGY CAPACITY FOR ELECTRONIC PORTFOLIOS

5 – Hurray! You are ready to go! (Of course there is always bigger and better!)
> All equipment outlined in level 4 plus: Full TCP/IP Internet access with a T-1 line or cable modem from district server. Rewritable CD-ROM, two more networked computers with 64 + megabytes of RAM, digital video, and audio editing hardware and software. Extra gigabytes of storage (such as JAZ drive). Storage capabilities on a school or district server.

4 – You are in great shape! Just keep your eyes open for any new equipment.
> Three to four networked computers, Ethernet network with 56K access from district server, 32+ megabytes of RAM, 2+ gigabytes of hard drive space, AV input and output, scanner, digital camera, high-density disk drive (Zip).

3 – You can make do, but it is time to brainstorm.
> One or two networked computers with 16 megabytes of RAM, 500 + megabytes of hard drive space, digital camera or scanner. Dial-up access to the Internet through a 28.8 modem.

2 – Don't let any grant writing opportunities pass you by.
> One single stand-alone computer, with 8 megabytes of RAM, 80 megabytes of hard drive space, no peripheral equipment, and 3-inch diskettes. Printer sharing and file sharing exists in parts of the building, just not convenient for student use.

1 – Oh, no! We have a problem! Need to locate some major funding.
> No computer.
> No network.

Figure 2.1

gration in Figure 1.1 as a guide (see chapter 1). Professional development activities can then be tailored to existing needs.

Professional development and equipment acquisition must be coordinated to occur at the same time. Otherwise, the training is forgotten if it is not put into practice, or the equipment stands idol because the educator does not know how to turn it on or open the application. After professional development sessions, teachers must continue to have access to resource persons so that they can make use of what they heard or tried during a presentation.

Following are several points to consider when implementing a professional development program:

1. Check on the arrival date and installation of equipment; coordinate training so both occur at approximately the same time.

2. Establish mentors for teachers and a process for contacting them after initial training.

3. Schedule training based on current level of expertise.
   Differentiate professional development according to the participants
   —content specialty, level to expertise with technology, and so forth.

4. Provide a session (hands-on) to understand the basics of the application. Relate application to how it can be used in the classroom,
   allowing teachers to create a product similar to one a student would
   create (see the Plugging in the Portfolio sample in this chapter).

5. Provide follow-up training linked to goals of the school or district.

6. Refer teachers to any online training available on the application
   they can access between sessions.

7. Schedule classroom observations to see how this knowledge is
   transferring to students.

8. Assess the effectiveness of the professional development, and make
   any necessary adjustments.

If equipment can be purchased and added to an existing system,
teachers can select from a number of options, including all the media
components—text, graphics, audio, and video. Hardware and software
decisions also depend on the skill level of students and teachers. An electronic portfolio system can be established using only text files created
with a word processing program. Such work is usually within the level of
expertise for most students and teachers, or they can be quickly brought
up to speed with practice and experimentation. As skills increase and
more equipment can be purchased, graphics, audio, and video can be
added, allowing for a more fully realized electronic portfolio.

Assessment

Access and storage issues, portability, and increased communication flexibility make electronic portfolios a valuable tool for assessing student
achievement. Integrating technology into assessment improves the quality and timeliness of the feedback process. Electronic files stored on disks
do not take up much physical space and can easily be opened away from
school. The Internet offers even more flexibility for access anytime, any-
where, even when a student wants to use the portfolio to conduct a conference though a parent lives or works across the nation or across the
world. These characteristics are key factors when assessing understanding if students are going to be required to demonstrate what they know
and can do as they proceed through their K–12 education.

As discussed, not everyone begins at the same level because of
equipment limitations or technology expertise. Many tools are currently

available with different advantages and disadvantages. Various types of popular software are discussed here in relation to a teacher's level of technological proficiency.

Database Software

Databases are large files that tend to be teacher-centered and difficult to use when creating written documents. Teachers need to have a high level of software understanding in order to successfully transfer its use to students.

Teachers whose expertise falls into the Adoption level can utilize database software with predesigned templates. At the Appropriate level teachers should be acquiring the skills necessary to develop relational databases. Database software makes the tasks of creating checklists, tracking student work, sorting standards, and compiling outcome relationships much easier. Most database applications are network friendly, available in both PC and Macintosh platforms and have multimedia capabilities.

Hypermedia "Card Format"

Teachers at the Adaptation level will find this software well organized with easy to understand tools to produce creative products. It is usually available in PC and Macintosh platforms, both with excellent multimedia capabilities, graphics, text, sound, animation and video. This type of software is somewhat Internet accessible, but it often needs a downloadable plug-in for the browser. Products take up a great deal of storage space, and it can be difficult to import information from other programs.

Multimedia Authoring

This type of software requires a great deal of time and effort to learn. Teachers at the Invention level will find flexibility when facilitating students' publishing because of the ability to include graphics, text, sound, navigational buttons, animation, and video.

Multimedia Slide Shows

This software gives teachers at the Adaptation level a starting point to integrate graphics, sound, or video files into student products. This presentation software is usually included in most office bundles.

PDF (Portable Document Formats) Documents

Teachers at the Adaptation level can create documents using other programs. They then save files in PDF format, making them Web-accessible with an Acrobat Reader plug-in downloadable from the Internet. At the

Appropriation level teachers edit files with an Acrobat Exchange. These files are larger than HTML (HyperText Markup Language) ones and may lose formatting when changing contexts into PDF. Due to the number of steps involved, they lend themselves to use with older students.

Video

Teachers at the Adoption level use analog video to support traditional instruction. Analog video costs less, but storage is bulky. This is a good starting point for integrating technology when sharing performances with parents. Digital video enables teachers at the Invention level to edit and organize video clips and upload to the Internet if the expensive digital video equipment is available.

Web-Based Authoring Tools

Teachers at the Adaptation level can facilitate students as they create text and graphics using HTML Web-publishing software. Teachers at the Appropriation level can facilitate older students without purchasing Web-publishing software. Students can create their portfolio by writing HTML code. Difficulties may arise with sound and video files. Security issues may present a challenge when publishing on the Web.

With all the choices on the market it is difficult to determine which product is the most suitable. Appendix B lists Web sites that teachers can visit to explore software options to fit their programs and pedagogy. Teachers might want to reference the glossary of technical terms at the back of this book as they read Figure 2.2, "Equipment and Storage Devices to Support an Electronic Portfolio Project."

Working with Existing Equipment

Each student has a story to tell. If teachers wait until the latest and greatest technology is placed in their classroom, many stories will be missed. If technology skills are limited, begin with a whole-class journal portfolio. Figure 2.3 is an example of what might appear when viewing the contents of a student folder created with a word processor. Reflective comments can be added at the end of the document. The student can open the file, scroll to the end, and begin adding comments. A rubric can be used to assess the journal portfolio. It can be placed at the end of the file or before the reflective comments. In this continuous journal format all information is located in one file. This can be a teacher-directed or student-directed activity.

EQUIPMENT AND STORAGE DEVICES TO SUPPORT AN ELECTRONIC PORTFOLIO PROJECT

Equipment

COMPUTER—a stand-alone or networked computer

SCANNER—a color flatbed scanner is the best choice

DIGITAL CAMERA (full motion video or still image)—close-focus adjustment to capture student writing samples

VIDEO CAPTURING HARDWARE—(captures and digitizes video from outside source, i.e., VCR tape)

Storage

HARD DRIVE on a stand alone computer

FLOPPY DISK—stores 1.44 megabytes of information or approximately 250 pages of text

ZIP DRIVE—stores 100 megabytes of information or approximately 1600 pages of text

JAZ DRIVE—stores 1 gigabytes of information or approximately 160,000 pages of text

COMPACT DISK (rewritable)—stores text, sound, graphics, and video, approximately 300,000 pages of text

DIGITAL VIDEO DISK (DVD) (rewritable)—stores text, sound, graphics, and video, approximately 2.6 to 17 gigabytes of information

WORLD WIDE WEB SERVER—Web based storage changes the format of graphics to take up less space than if stored on hard drive

VIDEO TAPE—stores video footage with or without a computer to organize information

AUDIO TAPE—stores audio recording with or without a computer to organize information

Figure 2.2

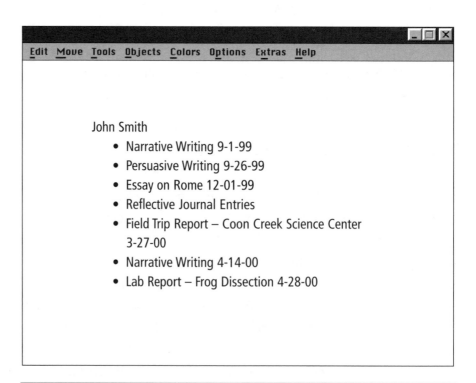

Figure 2.3

An electronic portfolio project can even begin with one stand-alone computer and a word processing program enabling students to generate and manipulate text. Technology reduces time spent on prewriting processes, drafts, and creating template files. This allows more time for writing and focusing on content. Brainstorming is made easier with visual mapping software that helps students clarify thinking, organize thoughts, and make connections between ideas (see Figure 2.4). Many times brainstorming is overlooked because of the lack of time. Efficiency in creating an idea map increases the likelihood that this step is included to generate multiple ranges of possibilities for topics, goals, performances, and assessment processes.

MAPPING SOFTWARE

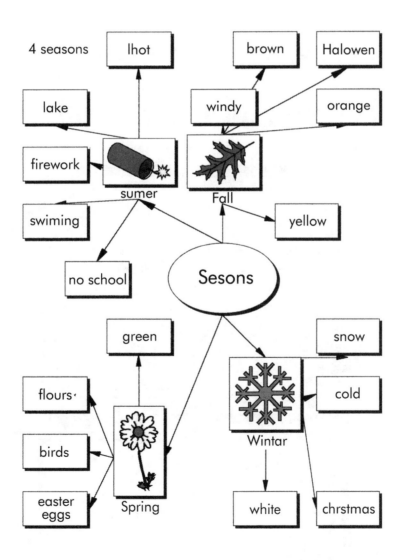

Figure 2.4

Students can create writing samples, reports, essays, scoring criteria, standards, rubrics, and self-reflections while meeting standards and demonstrating understanding. Various forms of work (video, audio, text, and illustrations) can be stored electronically on the hard drive of the classroom computer or on a 3-inch diskette. Another storage device (CD, Zip, Jaz, or DVD) may be needed to increase storage space. Digital images can also be imported into presentation software just like clipart files.

When storing work on the hard drive, a folder can be created for a student, and all that student's work can be saved in individual files within his or her own folder. Saving work on individual diskettes increases portability and decreases the chance a whole portfolio could be lost if a diskette were misplaced. Another advantage of storing work on diskettes

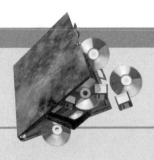

Plugging in the Portfolio

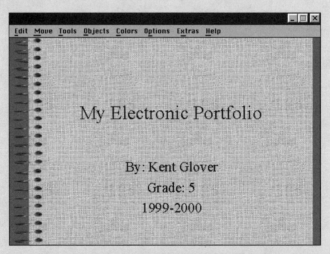

SAMPLE 2.1 This is a cover page created by the student using a teacher-made template. Microsoft PowerPoint software was used. Using the down arrow key on the keyboard, one can move to the next slide.

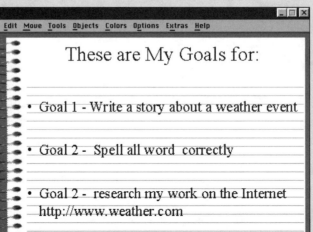

SAMPLE 2.2 Using Microsoft Powerpoint software the student was able to complete the screen from the teacher-made template. Using the down arrow key on the keyboard, one can move to the next slide.

is that students can then add to their portfolios from any computer with the same word processing software. If a student began a narrative writing piece in Microsoft Word 2.0 and wanted to complete the assignment for homework, the work could be saved on a diskette and continued after school at the library or at home as long as both computers were the same platform. (Platform refers to whether the machine is a Macintosh or PC.) It is important to note that files created in an earlier version of a software application can always be opened with a later version, but an earlier version cannot open files created in a later version. Macintosh computers can save to PC formatted diskettes; however, PC computers cannot save or open Macintosh diskettes.

Portfolios can be thought of as filing cabinets. Separate files can be created with the cabinet (portfolio) on a diskette, CD, or hard drive for

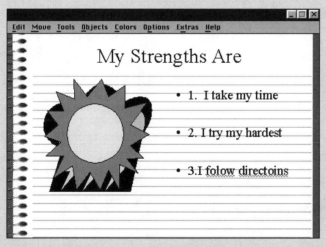

SAMPLE 2.3 The student uses the template as a place to record his reflections upon his personal strengths. Using the down arrow key on the keyboard, one can move to the next slide.

SKILL LEVEL

- [ ] Entry
- [x] Adoption
- [ ] Adaptation
- [ ] Appropriation
- [ ] Invention

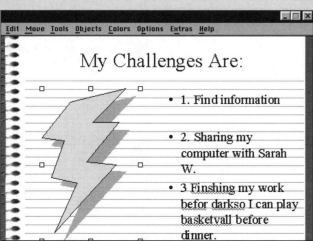

SAMPLE 2.4 Still using the template, the student is prompted to record his reflections. Using the down arrow key on the keyboard, one can move to the next slide.

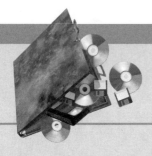

Tips to Remember

- If software does not automatically record a date on the document, make a note within the work.
- Label files, directories, and folders for easy access.
- Save work frequently.
- Remember to add page numbers.
- Put name on all work added to networked files to identify who contributed.
- Plan and outline work prior to computer time to avoid wasting time.

each draft of a writing assignment. Separate folders can be created for each student, and within those folders a separate file can be created for each subject as well as one for reflective writings. Always update and back up files on the hard drive, if space is available, in case the diskette breaks or is lost.

Word Processing and Spreadsheets

Many school districts purchase a suite of software in one bundle containing a word processing program, spreadsheet, and database, such as Microsoft Works or Microsoft Office. Students can graph information using a spreadsheet program or categorize data using a database program and save their work in the same folder on the hard drive or on their diskette regardless of the program used. When viewing files be sure to select "View All Files," otherwise only selected files in a particular format will appear.

Plugging in the Portfolio

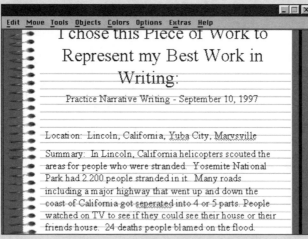

SAMPLE 2.5 This sample represents a piece of work the student has chosen as his best work. The student placed the paragraph directly into the teacher-made template. Using the down arrow key on the keyboard, the student can move to the next slide.

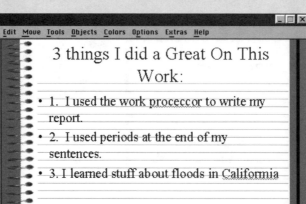

SAMPLE 2.6 This PowerPoint slide offers the student's justification for why he chose the work in sample 2.5 as his best.

Multimedia Software and Devices

Still using only a stand-alone computer, but adding multimedia software such as Hyperstudio or Mi#rosoft PowerPoint, students are able to present their portfolio in a slide show format. A series of cards or screens are linked together to form a presentation stack. This format lends itself to a project, individual, or group event, or to a thematic portfolio where students are telling a story. Text can be typed directly on the card or copied and pasted from other files. Pictures can be added from a bank of clip art images without any additional equipment.

Digital Cameras

If a digital camera is available, students can take a picture of an image and download it directly into the computer. Digital cameras operate like traditional cameras except pictures are in the form of digital images instead of developed on paper. This piece of equipment could be used to

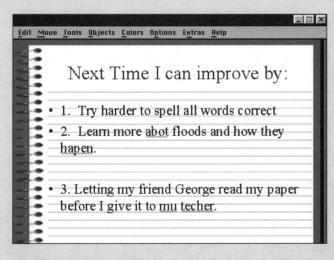

SAMPLE 2.7 This page or slide in the template gives the student the opportunity to reflect upon his or her performance and determine three strategies for improvement to implement during the next assignment.

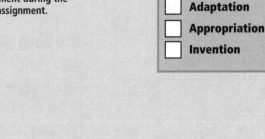

SAMPLE 2.8 The teacher completed the Writing Rubric when the template was developed. The rubric is included to let the student know how the writing piece would be scored.

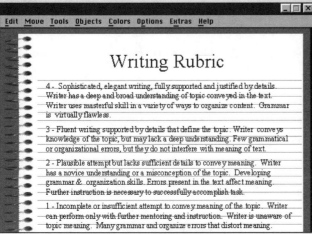

quickly capture an event to later add to one's portfolio. For example, if a class visited the zoo, the digital camera could be used to take pictures documenting the trip. Upon return to the classroom, a "Zoo Trip" folder can be created on the computer desktop and pictures downloaded as individual files into the folder. Students later open the files, sequencing and labeling each to tell their zoo story. Remember that picture files are large and may not fit on a 3-1/2 inch diskette if storing on the hard drive is not an option. Storing digital images on a CD may be an option if the appropriate equipment is available.

Scanners

Scanners can be used to capture text, authentic pictures, or graphics originating on a piece of paper. Scanners read the patterns of light and dark on the paper and store them as digital information. Like the digital camera, scanned images can be inserted in other documents as one would with clip art. The advantage of scanned images is that they can be altered—made lighter or darker or cropped then saved in several standard formats (i.e., bitmap, JPEG, TIF, GIF). All can be imported into most software programs as graphic files.

All scanners come with standard imaging software, which allow teachers to take a digital snapshot of a document. If a text file is scanned and needs editing, optical recognition software (ORC) is required. ORC software enables one to edit text within an image instead of just resizing it. Images are converted into letters and can be opened and edited with word processing software. Many current OCR programs can also store formatting information.

Moving to the Next Step

So far, only word processing software and stand-alone computers have been discussed. With this level of implementation, artwork, 3-D models, videotapes, and audio tapes would still need to be stored in a traditional way until additional multimedia options are available. Teachers at the Entry or Adoption level of expertise can implement electronic portfolios with word processing software and stand-alone computers. If artwork, 3-D models, videotapes, or audio artifacts are to be included, they would still need to be stored in a another format until additional multimedia options are examined and professional development has been undertaken. At a later point, these artifacts could then be digitized and added to the electronic portfolio. Expanding into multimedia increases the types of files that can be included to meet project needs. Typical artifacts students may add to a portfolio over a period of time might include the following:

- Goals and learning outcomes established by the student and/or teacher prior to beginning a project or at the beginning of the year
- Standard(s) that measure what a student understands to ensure all aspects of the teaching and learning process have occurred based on the student's performance
- Student work samples demonstrating what a student knows and is able to do
- Student reflections that ensure students assume responsibility for monitoring the quality of their work and establish clear goals prior to beginning the project
- Teacher and parent assessment and feedback to strengthen the partnership among students, parents, and teachers

The Venn diagram in Figure 2.5 illustrates the advantages and disadvantages of electronic and traditional manila folder portfolios as well as listing portfolio characteristics. As schools prepare students for a job market driven by microchips and fiber optics, they cannot afford to only teach the old basics of reading, writing, and arithmetic. New skills and competencies must be in the forethought of education.

In 1991, the US Department of Labor released a landmark study by the Secretary's Commission on Achieving Necessary Skills (SCANS). This Commission, consisting of a group of prominent business, labor, education, and government leaders, set out to identify the changes occurring in the workplace. They determined that a traditional skillset was no longer sufficient. High-paying assembly-line jobs no longer exist to the extent that they once did. These jobs have been largely replaced with positions that require workers to be competent in computer literacy, problem solving, interpersonal skills, goal setting, working as part of a team, and allocating resources.

The skills identified in the SCANS report are the same ones needed when electronic portfolios become part of a student's learning experience. They are a means for educators to equip students with the critical skills employers expect of workers in a high performance workplace. The skills include the following:

- Outcomes/learning goals
- Standards/rubrics/assessment criteria
- Student work samples
- Student self-reflections
- Teacher assessment and feedback

These options ensure that students are not just creating electronic file cabinets, but can effectively display what they know, what they can do, and what they understand.

Networked Technology

When working on a Local Area Network (LAN) or Intranet, files can be shared throughout the building. Technology (computers and other supporting equipment) can "talk" to each other, enabling student work files and/or the software used to create them to reside in a central location. Networked versions of software are loaded on a server allowing student access from any workstation on the network. The number of students using the same networked software application simultaneously depends on the number of users specified in the software license.

CHARACTERISTICS, ADVANTAGES, AND DISADVANTAGES OF PORTFOLIOS

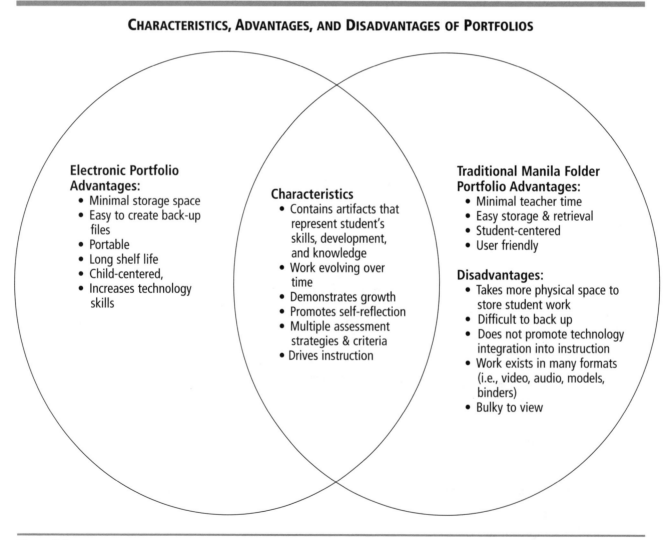

Electronic Portfolio Advantages:
- Minimal storage space
- Easy to create back-up files
- Portable
- Long shelf life
- Child-centered,
- Increases technology skills

Characteristics
- Contains artifacts that represent student's skills, development, and knowledge
- Work evolving over time
- Demonstrates growth
- Promotes self-reflection
- Multiple assessment strategies & criteria
- Drives instruction

Traditional Manila Folder Portfolio Advantages:
- Minimal teacher time
- Easy storage & retrieval
- Student-centered
- User friendly

Disadvantages:
- Takes more physical space to store student work
- Difficult to back up
- Does not promote technology integration into instruction
- Work exists in many formats (i.e., video, audio, models, binders)
- Bulky to view

Figure 2.5

Networked computers increase flexibility for students to work from multiple locations throughout the building or campus. As teachers schedule time for students to use computers, discussion with colleagues is needed to decide when students are able to use equipment. The media specialist in the library may set blocks of time during each week for a small group of students to work independently. Classroom, science lab, music, and/or computer lab teachers (any staff member who may have open time on their computers) may open up blocks of time for students from other classes to schedule time in the room. The key to maximizing productivity is to communicate with colleagues and carefully plan ahead of time.

If the culture throughout the school is already allowing students to move about the building, adding electronic portfolios will not be difficult. However, if electronic portfolios introduce a new concept in the way students work, teachers should meet with the principal and/or curriculum team to address specific needs for implementing the project.

Safety and Ethics

As schools meet the challenge the White House presents to connect every classroom to the Internet, educators will increasingly incorporate this new tool into electronic portfolio projects. Use of the Internet raises new questions as more and more students will be placing personal information into the electronic world. Parents and teachers are concerned with student safety on the information highway, access to inappropriate sites, and communication with strangers. The US Federal Trade Commission found that 89 percent of 212 children's sites surveyed in June 1998 collected personal information from minors. Fewer than 10 percent had any parent control over what was being collected (Teicher 1999).

Teaching students to use the Internet responsibly and safely will ensure its success in classrooms around the world. Protocol for publishing work on the Internet should be established, especially if it is not password protected. In addition, it is important to keep the following in mind:

- Never post personal information (last name) so that the students can be located by strangers.
- Establish where, when, and how Internet sites can be accessed.
- Maintain ongoing discussions regarding the students' online experience.
- All students' must have a current "acceptable use" form signed by their parent or guardian on file at the school.

- Teachers should continuously monitor students as they work on the Internet. It is important to arrange the classroom so it is easy to view the computer screens.

Ethical behavior in cyberspace is another subject for inclass discussion before students begin researching and publishing their electronic portfolios on the Internet. Topics such as copyright, copying and distributing software, privacy, hacking, and obscenity can be discussed in the context of basic values and the importance of truthfulness, responsibility, and respect for others. The following are important points that should be communicated to students:

- Only when the source of information grants permission to download a software program from an Internet site is it permissible to do so; otherwise, it is considered stealing.
- To many children the Internet may feel like a pretend world since it is not concrete and cannot actually be touched. It is appropriate to create a special online name to keep one's anonymity, but pretending to be something or someone to deliberately deceive another is not acceptable.
- Permission must be obtained from the owner to copy photographs, music, stories, films, or other forms of artistic work. Some pictures or images are considered public domain and can be used freely. With e-mail capabilities, obtaining permission has been made easier.
- Software is usually copyrighted. Some programmers create "freeware," meaning it can be copied or shared for free. Some software is considered "shareware" or "demo" and can be downloaded legally and used for free for a limited time period.
- If e-mail capabilities are added to a student portfolio it generally is considered private if the correspondence is between two individuals. Due to the ease with which one can misrepresent him or herself, students should not provide any additional identifying information online unless they are confident who they are replying to and how the receiver is going to use the information.
- When students are exploring the Internet they must not attempt to access a private information system requiring an ID and password to enter. Students should exit that area of the Internet.
- Information considered obscene or objectionable by some may still be protected under the First Amendment of the Constitution and may not be legally obscene. Under the current law, whatever is considered legal in print is usually legal on the Internet.

Such issues are on the agendas of many technology coordinators and school boards across the country. Some have put in filtering software to attempt to solve this problem. The disadvantage to this approach is that many educationally sound sites are unavailable to students because words are present that the filter screens out. For example, a 10th grade biology class may be studying the human body and searching for information on breast cancer. A filtering software program could associate the word "breast" with an obscene site and filter it out.

Connectivity to the Internet is rapidly becoming a reality in classrooms across the nation. Portfolios stored on the Web can be viewed from any computer in the world, regardless of platform, as long as it is connected to the Internet. As students move through their K–12 education, Web-based portfolios can easily follow them since the portfolios can be stored in a central location.

Whether the school is connected to the Internet in every classroom, in one location, or only within the building on an Intranet network, consult with the technology coordinator or person responsible for maintaining the network. This person will be a key resource when determining existing equipment capacities, making purchase recommendations, and providing insight into strategic planning for beginning or expanding this or any other technology-based project. Teachers will always want to consult with their school's technical support team before starting the project.

BEST PRACTICES IN AN ELECTRONIC ENVIRONMENT

The Instruction and Assessment Connection

High quality assignments require students to solve problems, read with comprehension, and write with fluency to produce a product. These are the same types of products and performances all portfolios should contain. The advantage in creating electronic portfolios is that the very skills needed for this type of project are the same ones employers will require in the high-performance workplace of the twenty-first century. Implementing electronic portfolios as an instructional tool is a powerful way to help teachers, students, and parents identify instructional strategies, types of tasks, intelligence, and learning processes that work best for that particular student. When evidence of learning can be organized in an electronic portfolio and presented to the audience by the student, then clear goals, objectives, and standards can be determined for continued student growth and achievement.

The collection, selection, reflection, and projection processes requisite in a portfolio system change the whole classroom environment. Invite parents to participate in these processes, and give them the opportunity to see samples of their child's work and know what learning is taking place. One strategy to involve parents is to periodically send the collection of work home and encourage parents to write comments about it. Some teachers prefer a "Portfolio Night" at the school when students share work with their parents. The contents of electronic portfolios may also

Plugging in the Portfolio

The sample portfolio in this chapter represents one created when teachers are at the Adaptation level of expertise. At this level teachers encourage *all* students to utilize technology in classroom practices. Technology is being integrated into the curriculum; however, instruction is still traditional in nature. The individual student portfolio shown in this chapter is student-centered with the teacher coaching the activity. It incorporates several different software applications. The portfolio is created using Hyperstudio presentation software with images imported into the stack. Teachers are at an emergent level of managing groups of students using technology; therefore, only one group activity was attempted and included in the student's portfolio. Other software applications represented in this sample are Inspiration, Microsoft PowerPoint, and word processing. Other work was scanned into an electronic form and imported into the portfolio.

be transmitted as attachments to e-mail, which allows parents with access to the Internet to view the material and respond electronically. Any method chosen has the potential to increase communication among parents, teachers, and students—an important goal.

Creating a traditional portfolio does not require student work to be in a particular format, it just needs to fit in the box. On the other hand, when students create electronic portfolios, all work is eventually digitized. Whenever possible, students should be encouraged to create their work using technology (word processing, digital camera, concept map software). This eliminates the step of transforming paper and pencil work into a digital format, which makes the selection stage more manageable. Prior to assigning any task, the teacher should ask him or herself, "How can this student work be created with the least number of steps to eventually be added to an electronic portfolio?" The more work that can be *created* with technology, the less that needs to be digitized format. Working directly in an electronic environment may not always be feasible, but it pays to use technology in the creation stage whenever possible.

Following is a guide for digitizing several different types of student work by product (artifact):

Product: 2-D work
Examples: Essay, drawing, photograph
Plugging it in: Take processed film and scan it into the computer or, better yet, use a digital camera to take the picture.
File Type: GIF files, Bitmap, JPEG, and TIFF

Product: Audio recording
Examples: Poetry reading by student, speech made in drama class. Some audio files can be downloaded from the Internet to support a report (e.g., "I Have A Dream Speech" by the Rev. Martin Luther King, Jr., or President Roosevelt's Pearl Harbor speech). If the site states the work is public domain then it can be used; otherwise, permission must to obtained.
Plugging it in: Audio recordings can be digitized using any Wave form recording software. (Some multimedia software has audio recording capabilities.)
File Type: WAV, MP3

Product: Video recording
Examples: Student performance in a play, dance, or debate
Plugging it in: Video recordings can be created in digital or analog

format. Digitized video can be captured with a digital video camera and directly imported into the computer. Analog video is captured with a video camcorder using a VCR tape. The VCR tape is then converting into a digital format using a video capture card. Other digital video editing software include Avid Cinema and Adobe Premiere. Some multimedia software has video capturing capabilities. Note: Some of these video-editing programs require an Invention level of technology expertise.

File Type: MPG, AVI, MOV (QuickTime).

The Right Stuff

Professional development is needed for teachers to navigate through an all-inclusive, multimedia software application. Teachers at the Entry or Adoption levels may feel overwhelmed with all of the construction and peripheral equipment options available. Teachers who are at the Adaptation level are beginning to integrate technology across the curriculum. They may find commercially produced portfolio software (multimedia) easy to use and all-inclusive in helping to organize the project and eliminate some of the unknown design factors. Several programs such as Hyperstudio, Scholastic's Electronic Portfolio, or Grady's Profile allow students to blend various media types in a single file, including tools to constructs graphics, sound, animation, and possibly movies.

Organization

No matter how in-depth or comprehensive a portfolio, it is how the work is displayed that makes the strongest impression. Electronic portfolios present students' work in an organized, sequential manner that provides concrete evidence of what the student knows, understands, and can do. The way a portfolio is organized can make it easier or more difficult to determine student achievement. There are several ways to organize electronic portfolios, including the following:

Chronology

This most common form of organization displays student work by date. It is not always as interesting an arrangement because students' best work might be lost or mixed throughout the collection. The potential audience for this type of portfolio can include the principal, guidance counselor, student's teacher at next level, promotion or graduation committee, school psychologist, and parents. An example of a portfolio organized chronologically can be found in chapter 1.

Academic Discipline

This classification by subject shows strengths in various fields, including writing across the curriculum. If a student is strong in mathematics, but weak in language arts, artifacts found in the portfolio will make that evident. Satisfactory achievement of a writing standard may not be shown in the student's language arts work but may be present when the student uses language concepts in a mathematical context, allowing the student to show what he or she knows instead of what he or she does not know. The audience to which the portfolio is directed is the same as it is for the chronologically organized portfolio. A portfolio organized along academic discipline lines is shown in chapter 4.

Genre

Organizing by genre gives flexibility and clarity to the portfolio. Work can be compared to show growth in a certain area. For example, similar writings (narratives, debates, persuasive essays) can be easily compared regardless of the content. Chapter 2 provides an example of a portfolio in the science genre.

Draft to Final Product

This structure is the best way for an observer to gain insight into the process the student went through to reach the final product. If reteaching or extra practice is needed in a particular area, it is easy to pinpoint. The planning, revising, and other steps taken toward the final product can all be shown. The drawback with this organizational structure is that it can become cluttered and difficult for an observer to wade through the work. Chapter 4 shows an example. The likely audience for this kind of portfolio includes parents and teachers.

Worst to Best Work

This organizational structure is the best way to show growth over time. It is especially helpful if the student's reflection is attached to critique the work. An example of this kind of portfolio can be found in chapter 5. Teachers, parents, guidance counselor, and the principal can learn much about a students' progress from viewing a portfolio organized in this manner.

Individual versus Group Work

Skills needed in the workforce, such as teamwork and responsibility, as well as individual accomplishments can be displayed in this structure. If individual student portfolios are stored on the Internet, a link could be made to a class portfolio to display group work. Other electronic storage methods (CD, diskette, Zip disk) can have a reference to work stored in

another location. This can be noted in a Table of Contents file or identified in a concept map that categorizes the contents of the portfolio. In
addition to educators and parents, potential employers may view this
type of portfolio. See chapter 3 for an example.

Theme

Teachers are integrating curriculum and construct interdisciplinary units
to teach beyond the facts in order for students to gain a deeper understanding of the subject matter. Organizing student portfolios by theme
can be an effective and easy way for the audience to understand what
knowledge the student gained as a result of the study. The principal, parents, teachers from other grade levels who may be teaching the same
unit or theme, and the student's peers may be among those who would
view this type of portfolio. An example of a thematic portfolio can be
found in chapter 3.

Concept Mapping

This visual arrangement gives a cumulative picture of the student. A central idea is established to provide a focus for the portfolio. Work supporting this central idea is included along with a reflection statement to clarify support. In an electronic format, this organizational structure is easily
achieved with software such as Inspiration or through a table of contents
file that identifies all work supporting the central focus. Chapter 4 offers
an example.

Content Standards, Goals, or Objectives

When goals are established in the beginning of a course or at the beginning of the school year, work can be arranged in a portfolio to demonstrate progress toward or achievement of standards, goals, or objectives.
See chapter 5 for an example.

Collecting, Selecting, Reflecting, and Projecting

Many of the same decisions made when constructing traditional portfolios carry over to an electronic format. Once the portfolio's purpose and
audience is established, defining the process for collection, selection,
reflection, and projection—the essential steps critical to the organizational flow—are followed.

1. **Collect** everything for a working portfolio.
2. **Select** pieces for the final portfolio.
3. **Reflect** on selections.
4. **Project** future goals and how they can be accomplished.

Collection

The ongoing process of collecting student work requires planning and organization on the part of the teacher *and* the student. As teachers return scored work to students, they should decide which pieces are appropriate to add to student portfolios. Digitizing authentic work that demonstrates standards and goals can then be scanned into the computer. This is an ongoing process; an efficient system must be set up to accomplish these tasks in a timely fashion.

Managing what students add to their electronic portfolio on a continual basis is achieved by setting up a system. For example, teachers can place student work they have scored in a hanging folder with a note on it recommending it be considered as a possible portfolio entry. On a regular basis, students can look through the contents of their hanging folders to make portfolio selections.

Teachers may want to structure the selection process with specific criteria to guide the student. A template could be made on the computer containing the information (see Figure 3.1). Using the template on the computer, the student can fill out the questions on the form to complete their reflection of the work, or they could have a file in a folder on the desktop of the computer and add a reflective journal entry. If work is already digitized (e.g., word processing, spreadsheet), the teacher could place a note in the student's hanging folder or send the student an e-mail message suggesting a digitized piece of work for inclusion in the portfolio. Otherwise, students may keep track of work as they begin collecting it by using index cards (see Figure 3.2) in a card box or clipped to the actual artifact. Or, they may use a log sheet for easy reference (see Figure 3.3), depending upon what works best in the class. Index cards and log sheets can be arranged by subject, specific skill, or date until the selection process begins. If a student removes an artifact, he or she should make a note giving the reason for removal.

Remembering a few points in the collection stage will make the selection process easier. In the collection stage, do *not* save everything. Work given to students for skill practice or to recall facts does not demonstrate deep understanding. Students still need to engage in fact-based learning. They need to gain knowledge and practice process skills that require them to remember and recall information before they can apply it to new situations. These types of activities may best be accomplished with worksheets and dittos, but they do not make good choices

STUDENT WORK REFLECTION FORM

Name:_____ Date:_____ Grade:_____

Standards addressed in work:_____

Task:_____

Who is involved in the selection process for this piece of work?_____

_____

What was learned as a result of this work?_____

_____

Why should this artifact be included in your portfolio?_____

_____

When was the work completed and how long did the task take to finish?_____

_____

How could the work be improved?_____

Signature:_____

Figure 3.1

ARTIFACT INFORMATION CARD

Name   Teacher

Grade   Subject

Title of artifact:

Date artifact was created:

Student's reason(s) for possible inclusion in portfolio:

(Optional) Scoring criteria:

Additional comments:

Figure 3.2

STUDENT PORTFOLIO COLLECTION LOG

Name:_____Grade:_____

Teacher:_____Date:_____

Title:_____Subject:_____

Individual/Group Work:_____

Reasons for Possible Inclusion:_____

Figure 3.3

for portfolios. Collecting work that represents specific criteria in the student's assessment plan is discussed in more detail in chapter 5.

Selection

Teachers can guide students' selection processes by providing quality, authentic tasks for them to do. The types of assignments that lend themselves for inclusion in portfolios are characterized by the following:

- Result in a product or performance important to students
- Are standards-based; students know what is expected of them prior to beginning
- Give students multiple opportunities to revise work
- Promote choices for accomplishing the task
- Are authentic, have meaning, and are valued by teachers, parents, and peers
- Lead students toward learning rich and relevant content

Plugging in the Portfolio

SKILL LEVEL

- [ ] Entry
- [ ] Adoption
- [X] Adaptation
- [ ] Appropriation
- [ ] Invention

SAMPLE 3.1 Students use Hyperstudio software to construct an electronic portfolio. Information is placed directly on slides or can be created in other applications and imported into the stack. Choices of fonts and how the navigation button is designed are left to students as they create the portfolio.

Designing authentic tasks requires teachers to consider the purpose and focus of the students' portfolio, which standard(s) they are working toward, and goals establishing for learning. Following is a sample of artifacts as they relate to the purpose of a portfolio:

Purpose of the Electronic Portfolio: Demonstrate Process Skills

Examples of Electronic Portfolio Artifacts by Skill

- Reading—Audio tape of student reading a passage
- Writing—Essay created with a word processing program
- Mathematical thinking—Written explanation of steps used in solving mathematical problem
- Speaking—Videotape of a speech or debate
- Listening—Audiotape, videotape, storyboard, or written explanation of the plot of a movie or story read aloud.
- Computing—Multimedia presentation of a unit of study

Different artifacts may be appropriate for inclusion in a portfolio with a different purpose.

Standards

The teacher should design lessons and units based on the standards document used by his or her school. Determining which standards and criteria to focus on ensures scope and sequence in the curriculum and that all standards are covered at each grade level. Therefore, material selected for inclusion in an electronic portfolio should reflect specific standards and identify the standards the work is meant to address. When teachers recommend a piece of work for consideration in a student's portfolio, there should be a systematic process to ensure standards, curriculum materials, and instructional strategies are aligned. Student work represents the desired knowledge and skills for a particular benchmark.

No matter what standards document a teacher is working from, there will undoubtedly be many standards that pertain to work at his or her level. Keeping track of what work is done to satisfy achievement of a standard is a daunting task, especially when this record keeping is done on paper. Several software programs not only enhance the organization and presentation of information but also help keep track of material. Database software helps students to structure, record, sort, organize, and investigate information. Technology reduces the time students spend on the routine tasks of setting up the information and allows for more time to analyze, evaluate, and construct meaning for the data. Creating statistical products using technology facilitates inclusion in an electronic portfolio.

Assessment by the teacher and self-assessment by the student drives future instruction. Teachers know what skills and concepts need further examination. Students know what they can include in their electronic portfolios to demonstrate their level of understanding of a content standard.

Reflection

According to Niguidula (1997), a critical factor in making the electronic portfolio a tool for reform is for students to take their work seriously enough to reflect on it and determine how they can make changes to deepen their understanding. When self-reflections are included in an electronic format, along with quality work, the educational system is truly changing for the better.

The *reflection,* or *metacognition,* stage is what makes portfolio classrooms different from traditional ones. In a traditional classroom, teachers grade all of the student work, and then return it. Most of the time it ends up in the "Bermuda Triangle of Completed Assignments." When portfolios are used, teachers still grade student work, but students place that work in their collection file. If the assignment merits possible selection for their portfolio, they engage in meaningful thought about their performance on that assignment. Students may choose to revise their work after reviewing the criteria given to them in a rubric. Student reflection upon their work strengthens their metacognition and ability to self-assess (Burke, Fogarty, and Belgrad 1994).

Reflection is often a new process for students. Teachers must model reflective behavior for students to know how to reflect upon their own work. The reflection stage engages students in the process of thinking about their learning in a very systemic way. An effective strategy for teaching reflection is to make a transparency of similar work from a student in another class or from a previous year. Divide students in small groups (three or four students) and have them score the work collaboratively. Once groups have finished scoring, offer open-ended prompts that lead in the reflection process. Examples include the following:

- I chose to describe this character in *story title* because I . . .
- I think I learned more about_____by using technology to . . .
- This is one of my favorite pieces of writing because . . .
- One thing I still need more practice on is . . .
- This project/unit demonstrates my ability to . . .
- I had difficulties with _____, but now I . . .

Encourage students to keep responses short; one insightful comment on each artifact is manageable. If space allows, comments could be

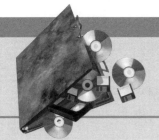

Plugging in the Portfolio

SKILL LEVEL

☐ **Entry**
☐ **Adoption**
☒ **Adaptation**
☐ **Appropriation**
☐ **Invention**

Individual "Working"
Portfolio Requirements

"Project Focus Folder"
1st semester
1. One Integrated Nature Trail project activity .
2. Tasks to represent math, science, literature, social studies and the arts.
3. Self & Peer Reflective Summaries
4. Rubrics or scoring criteria

"Project Focus Folder"
2nd semester:
1. One Integrated project example from Nature Trail.
2. Tasks to represent math, science, literature, social studies and the arts.
3. Self & Peer Reflective Summaries
4. Rubrics or scoring criteria

Requirement for Class Portfolio on Web:
1. One published writing
2. Published results from Group work.

"Special Pieces Folder"
1. These academic subject, one piece of work per grading period.
2. Selected by child/teacher
3. Self/Teacher/Peer Reflection Papers
4. Rubric or Scoring Criteria

Individual STUDENT PORTFOLIO

"Working" Electronic Folder Contained in plastic storage box on shelf in classroom.

From these folders, students' work can be selected for inclusion in Web based Class Portfolio

1. Audio cassette (record child's reading in the beginning of the year and again in the second semester).
2. Oral presentation or speech.
3. Computer disk with "working" files
4. Video tape of student performance.

Goals, Standards Vision Categories
1st & 2nd Semester - 2nd grade

1. Goal/Vision Category statements for each subject.
2. Achievement Test Skills & Subskills for grade level State Test

TIME LINE FOR STUDENT PORTFOLIOS
End of 1st semester "Project Focus Folder" must be completed
Portfolios must be partially ready for parent conferences (Late Winter)
MATH/SCIENCE/TECHNOLOGY NIGHT - Must be completed for this event with parents.

SAMPLE 3.2
Communicating the requirements of a task before the task begins ensures quality work is included in a variety of areas. The requirements are determined by the teacher and (in this case) imported as a GIF file into the student's electronic portfolio.

written at the bottom of the work or in a log. This will be especially helpful if the work must be digitized at a later date for inclusion in the electronic portfolio.

Each piece of work selected should be accompanied by a written reflection statement that details the purpose for inclusion. Young students should go through the reflection process even if they dictate their statements to an older student or adult.

Projection

The key to continuous improvement is strategic planning. It is important to plan the scope and sequence for all learning for the school year. However, students come into a classroom with diverse backgrounds and prior experiences. As students reflect upon their work in their electronic portfolios, they become more aware and responsible for their learning. Teachers can review reflection papers to gain a better understanding of what the student has mastered.

Plugging in the Portfolio

SKILL LEVEL

☐ Entry
☐ Adoption
☒ Adaptation
☐ Appropriation
☐ Invention

SAMPLE 3.3 Using Hyperstudio software the student creates a text box on one card in the stack to place text information. Buttons are place on the card to click on to move to the previous or next slide.

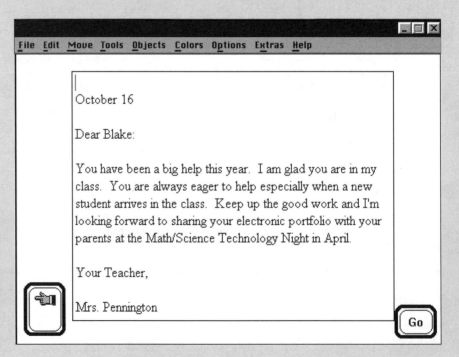

File Edit Move Tools Objects Colors Options Extras Help

October 16

Dear Blake:

You have been a big help this year. I am glad you are in my class. You are always eager to help especially when a new student arrives in the class. Keep up the good work and I'm looking forward to sharing your electronic portfolio with your parents at the Math/Science Technology Night in April.

Your Teacher,

Mrs. Pennington

Go

In the planning process teachers must look at standards, state testing objectives, and curriculum maps and then match those to the student's level of understanding to project what to do next. For example, a first grade teacher has engaged in "interactive writing" activities with a small group of students. During the class period the teacher models and demonstrates writing processes while involving the students as apprentices to work along side her. Students observe the conventions of print, spacing, punctuation, and organization of the words. Everyone has a chance to contribute and discuss how the story will develop to determine how all of the parts fit together to create a class book. The teacher selects words or other writing actions for individual students to do; the pencil is shared as the composition is created word by word. After completing the book, older students assist the group in scanning the pages to create a digital file to add to the electronic class portfolio. The teacher asks each student to reflect upon his or her work to project which students have the skills to move into "guided writing" activities.

Plugging in the Portfolio

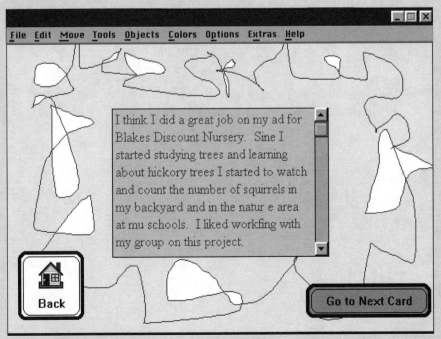

SKILL LEVEL

☐ Entry
☐ Adoption
☒ Adaptation
☐ Appropriation
☐ Invention

SAMPLE 3.4 Also using Hyperstudio software, the student can create a scrolled text box for placing written information.

Electronic Portfolio Outcomes

Creating student portfolios is an ongoing process. Throughout the project, one can look at the contents selected to determine student progress, patterns in work, and if achievement of benchmarks or goals are represented in the student work. One can also determine if holes exist in the selection of work (i.e., important steps in a project or unit, kinds of writing, particular subjects or period of time). This is especially important when the electronic portfolio layout and design stage begins so that a variety of higher-level thinking skills and a deep understanding of content is represented in the student work. The teacher should examine selections in the electronic portfolio to determine what needs to be added to ensure that all standards, objectives, and goals have been demonstrated in the work. The work included may also show that the student has not demonstrated achievement. But the tasks assigned by the teacher should provide the student with ample opportunity to show what he or she knows. From evaluating the contents of a portfolio, the teacher can gain the perspective necessary to drive future instruction.

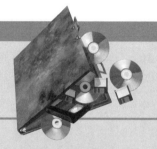

Plugging in the Portfolio

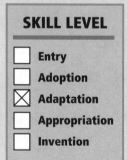

SKILL LEVEL

☐ Entry
☐ Adoption
☒ Adaptation
☐ Appropriation
☐ Invention

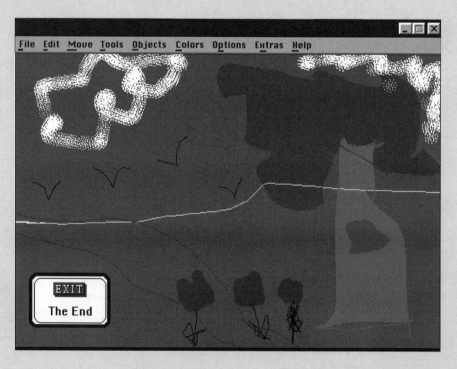

SAMPLE 3.5 Hyperstudio software also allows students to draw pictures and place a button to click on to return to the cover page.

Plugging in the Portfolio

SAMPLE 3.6 The student used a word processing program to print the text on paper, then used cut construction paper and markers to create a picture. Scanning the picture or using a digital camera to capture the picture in digital form enables the student to import the file into the electronic portfolio as an image file.

GETTING STARTED

How Best to Begin

As with other areas of pedagogy, strategies for implementing electronic portfolios are different for various grade levels. Younger children will require more guidance and whole group instruction, whereas high school students can be given longer blocks of time for independent work. Once the audience and purpose have been established, available technology has been matched to the project, and teachers have been striving toward *best practices* in authentic assessment, begin by asking what approach is best using the description of each of the following practices.

Teacher-Centered

Teachers are in control of the electronic portfolio design, layout, and construction. Parents may help, but students do not participate except for completing the work that goes into the portfolio.

Mixed Approach

Little or no student or peer participation in the collection, selection, or reflection processes. Teachers and students share responsibilities in one or more tasks: design, layout, or construction. Students lead parent conferences, collect some artifacts, help the teacher make final selections, and digitize some artifacts. Reflecting together and collaborating are good ideas.

Plugging in the Portfolio

At the Appropriation level teachers facilitate the use of technology as a tool to increase student learning. Learning experiences are designed to take advantage of technology to meet standards. Teachers give students choices to discover the best way to accomplish the task using technology as a tool to create their work products.

At this level teachers routinely integrate technology into all areas of the curriculum. This sample represents an individual high school student's portfolio. The portfolio sample is student-centered because it was created, designed, and constructed by the student.

Student-Centered

Students are in charge of their own electronic portfolio. They digitize their work, design, and layout and construct the actual portfolio. Students are responsible for self-reflections, gathering peer reflections and leading all parent conferences using their portfolio as a tool to demonstrate what they know, can do, and understand.

Questions to consider when deciding which approach works best with students include the age of the students, group dynamics, use of the portfolio, security, access, and student familiarity with technology. These characteristics are explored below.

What is the student age or grade level?

Younger students will need more mentoring as they create their electronic portfolios. Choose age-appropriate software and a storage device that can be easily used by younger students. Kid Pix software stored on a 3-1/2 inch disk will be more manageable for younger students than placing work on a Web-page software program such as Microsoft's FrontPage and storing it on the district's server. Pairing younger students with a buddy from an upper grade will ease classroom management.

Are students already working in cooperative groups to know what is expected of them?

Students need to be taught how to work within a group, including the different roles within the group and how to work with peers to accomplish tasks. Without structure and a systematic method for establishing cooperative groups, learning will not take place. Prior to beginning the project, teachers might want to research cooperative group learning. Books such as *Blueprints for Thinking in the Cooperative Classroom* (Bellanca and Fogarty 1991) can be of great help. Attending workshops or undertaking other professional development activities are also very instructive.

Are security issues a factor in the project?

As discussed in chapter 2, if portfolios are going to be stored on the Internet, there might be a safety issue if personal student information is included with them. The site may need to be password protected. Teachers should become familiar with district policy regarding safety and student use of the Internet.

Is the portfolio going to become part of the student's permanent record?

If so record, it is important that appropriate software be obtained to ensure that the portfolio material will be able to merge with other student records stored electronically. It is best to work with the district's information technology (IT) department regarding such matters.

Do students have a basic understanding of computer operation (including any peripheral equipment) and/or how to use the specified software?

If technological skills fall short of those required to implement a particular project, the content-area teacher may want to coordinate with the computer lab teacher to ensure that students are instructed in how to use and access equipment and programs to the extent that they can successfully work independently or in small groups. Engaging the expertise of the computer lab teacher will expedite technology training for students. In addition, if the knowledge and service of the computer lab teacher is available, it frees up the teacher to concentrate on the seemingly ever-expanding curriculum.

Creating a Project Plan

The following steps help teachers create a project plan and provide strategies for working with different grade levels. Adaptations can be made depending upon the student's age or individual situation.

Step One: Professional Preparation

Teachers begin learning how to use the necessary equipment, depending upon the their current level of technology understanding and adoption, as detailed in chapter 1. Professional development can be obtained by enrolling in a college class, finding a mentor, or enrolling in an online course to gain necessary knowledge of equipment, its connections, and how to install and use software. Teachers must have an understanding of this information in order to transfer this knowledge to students. Many online sources are designed to sharpen technology skills when and where it is most convenient for the teacher. Practice modules from basic desktop management show how to operate a digital camera (see Appendix B).

Step Two: Function and Form

Determine required contents for each portfolio based upon grade level, purpose, type, and audience. Although there is not one fool-proof method for choosing the purpose or type of portfolio that will work in a given circumstance, there are a few indicators that help focus decision making. When teachers begin to plan curriculum, instruction, and assessment, they look at the standards and specific expectations for their grade level. Correlating this information with what students are expected to know on the standardized test gives a good starting point. For example, if ten out of twelve science questions focus on life science in the fourth grade, and only two questions ask about physical science, it is obvious which types of science should be focused on in fourth grade.

Step Three: Organization

Create a storyboard or flowchart on paper to decide what information to include in the portfolio, order of entries, appearance, navigation of contents (linear or nonlinear), and organization. For younger children, the teacher should complete this task. A generic storyboard can be created and used with the whole class to manage the project. Upper elementary, middle, and high school students are quite capable of creating their own storyboards for their teacher's approval (see Figure 4.1). Once the teacher has approved the storyboard, students can begin a screen worksheet (see Figure 4.2). It is best if students only complete a worksheet for one screen at a time in case they need to modify or change information as they progress.

Step Four: Technology Skills

Choose the software and equipment needed for the electronic portfolio projects. Students should have an understanding of how to navigate through the software prior to beginning the project. Several options exist for teaching students the necessary technology skills.

Computer Lab Class

Coordinate with that support teacher to ensure students learn how to use the software chosen for their electronic portfolio projects while they are in the computer lab. If the portfolio project is school-wide, the computer teacher may need to meet with teachers as they plan in their clusters or grade levels to select software appropriate for each grade level.

Team teaching

If two or three teachers exchange students for different content areas, coordinate the teaching of the needed technology skills for all students involved. Determine what skills students already have and what needs to be taught. One teacher can become an expert in software and teach all students how to use the software. Another teacher can work on the selection/reflection process, while a third teacher can specialize in the layout of the work. Another teacher can teach cooperative group skills or how to use a scanner, digital camera, and so forth.

ELECTRONIC PORTFOLIO STORYBOARD

Use this form to design each page/card/slide of your electronic portfolio. Under each box, describe and make notes of any relevant details to organize the project.

Name_____ Date_____ Total # Pages/slides/cards_____

Comments:_____

Figure 4.1

SCREEN WORKSHEET

Basic information is outlined, allowing the student to organize his or her work for inclusion in the electronic portfolio.

Title Card

This is a personal cover page that should include student's name, year, teacher(s), picture, video or audio of student or written autobiography.

Figure 4.2

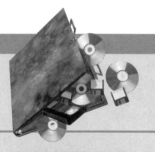

Plugging in the Portfolio

SKILL LEVEL

- [ ] Entry
- [ ] Adoption
- [ ] Adaptation
- [X] Appropriation
- [ ] Invention

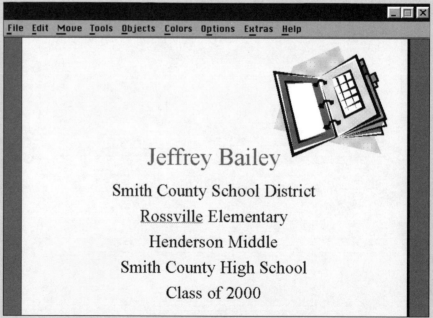

SAMPLE 4.1 This sample is a cover slide for a student's electronic portfolio used in a presentation by the student as one of the graduation requirements. The portfolio was constructed in Filemaker Pro. The cover page was created in Microsoft PowerPoint and imported into Filemaker Pro.

Self-contained classroom

If students remain in one classroom for all instruction and a computer lab class is not available, the teacher can teach technology skills to small groups of students on a rotating basis during center activities. Divide students into groups of four or five and have different tasks for each group. One center activity will consist of the "how to" skills needed for electronic portfolios. The teacher facilitates this center. Another approach is for the teacher to train a student from each group and then have those students mentor the others in their group.

Grade level or cluster teams

Teachers may teach in a self-contained environment but want to specialize in different areas of the electronic portfolio project. If several teachers are planning the same type of portfolio project, they may focus on specific skills and teach them to all students within the cluster or grade level, rotating teachers instead of students.

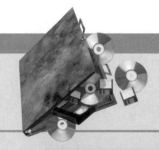

Plugging in the Portfolio

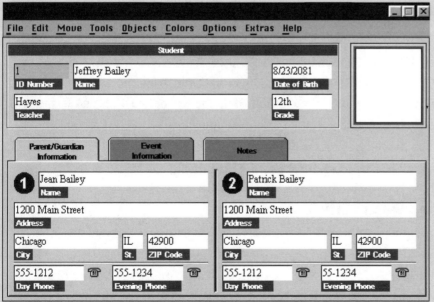

SAMPLE 4.2 The teacher can create a database of all student information and place each student's information into his or her respective electronic portfolio.

Departmentalized: Embedded within content areas

Middle and high school students usually have a different teacher for each subject, limiting the time spent with each. The advantage with older students is that most have had some exposure to software use and can pick up the "how-to" navigation skills quickly. Embedding electronic portfolios within content areas requires teachers to devote several class periods to navigate through the software and then pair students so they can mentor each other.

Departmentalized: Taught as a separate class

A computer class may be designed for the purpose of constructing student electronic portfolios. If this is the case, the teacher can present the purpose and concept of the electronic portfolios to colleagues, enabling them to provide students with tasks suitable for inclusion. Constructing electronic portfolios in a separate class and drawing upon the different academic areas for the content helps make connections between academic disciplines and utilizes technology as a tool to support learning.

Step Five: Responsibility for Learning

Teach students authentic assessment strategies giving them a strong foundation in the collecting, selecting, and reflecting processes to determine work for inclusion in their portfolio as discussed earlier in this book. Have students set up collection folders and begin placing work for possible inclusion into their electronic portfolios.

Step Six: Inclusion

Depending upon the age level of students, teachers can create a template to help them lay out their screen. Younger students, those with limited English proficiency, and those who are physically challenged may need to work on this task in a whole group setting while the teacher coaches and facilitates. Teachers may find it helpful to use a large white board and erasable markers to model the process. (Chapter 7 speaks more to special needs students and offers a sample portfolio to use with them.)

Step Seven: Efficiency

Examining the storyboard, screen worksheet, and/or design and layout form determine if any supporting work needs to be completed. Do pictures need to be taken, developed and/or digitized, an essay written, or an interview conducted? Teachers and/or students can develop a timeline, allowing time for each step and sequencing steps in the correct order. If students create timelines on their own, the teacher should approve the timelines prior to moving on to the next step. This ensures equipment is shared efficiently and equally, avoiding a situation where ten students are trying to use one computer at the same time.

Step Eight: Evaluation

Share and/or create a rubric with students to assess the electronic portfolio project. Many times work included in a portfolio has already been assessed. Whether the portfolio will be assessed as a whole or used as a display tool, students need to be aware of the criteria for creating an exemplary portfolio before they begin. Figure 4.3 is an example of a rubric that can be used or adapted to fit a particular group of students.

Step Nine: Time

Build time into the daily schedule, allowing students blocks of time for project work. Managing this kind of environment is discussed further in chapter 6.

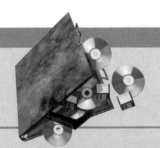

Plugging in the Portfolio

4th Grade Writing Assessment

SKILL LEVEL

- ☐ Entry
- ☐ Adoption
- ☐ Adaptation
- ☒ Appropriation
- ☐ Invention

SAMPLE 4.3 Many districts require students to demonstrate technology skills prior to receiving a diploma.

QUALITY WORK RUBIC

Scores/Criteria	Design	Knowledge	Organization	Quality
4 Exceptional	Work was meaningful, important, challenging, and engaging, reflecting real-world values. Understanding of concepts, guiding question clearly conveyed. Addresses standards.	Rich content, accessed data to construct meaning, clear integrated learning gained. Technology expanded knowledge base.	Order, structure, presentation shows direction, coherence and purpose. Connects to prior learning. Facilitates high level thinking and teamwork. Clearly relates to big idea and guiding question.	Real-world context, variety of resources used. Process skills evident throughout creating deeper meaning: reading, writing, listening, thinking, speaking, computing
3 Competent	Work was meaningful, important, and engaging, reflecting real-world values. Understanding of concepts, guiding questions is beginning. Addresses standards.	Accessed data to construct meaning, integrated learning gained. Technology expanded knowledge base.	Order, structure, presentation attempts to show coherence and purpose. Attempts to connect prior learning. Relates to big idea and guiding question. Teamwork does not necessarily support outcome.	Real-world context, variety of resources used. Process skills evident: reading, writing, listening, thinking, speaking, and computing; however outcome is at the knowledge level.
2 Emergent	Work lacks challenge for the grade level. Students are engaged, attempts to simulate a real-world connection. Understanding of concepts, guiding question was unclear, did not address standards.	Accessed data did not construct meaning, just facts. Technology used but ineffective.	Order, structure, presentation does not show coherence or purpose, but it does not affect meaning. Does not connect to prior learning. Big idea and guiding questions unclear. Teamwork does not support outcome.	Resources used inappropriately. Few process skills are evident: reading, writing, listening, thinking, speaking, computing, resulting in fact based outcomes.
1 Novice	Work lacks challenge for the grade level. Students were not engaged, did not address real-world issues. Understanding of concept, guiding question was lost. Did not address standards.	No clear knowledge gained. Integrated learning did not occur. Technology not used.	Order, structure, presentation does not show coherence or purpose, affecting meaning. No connection to prior learning. Big idea and guiding question unclear. No teamwork.	Resources not used. No process skills are evident: reading, writing, listening, thinking, speaking, computing, resulting in fact based outcome.

Figure 4.3

Step Ten: Reflection and Projection

Begin the project. Teachers can keep a diary or log throughout the project, making notes about what went well and what could be changed next time. This is especially important if the project is going to scale up in the future. These notes will prove invaluable!

Design and Layout

Once the teacher and/or students have the storyboard, screen worksheet, selected artifacts, reflections, design and layout sheets and are familiar with the criteria detailed in the rubric, it is time to arrange the contents. This process will vary depending on the software chosen for construction. Word processing software is limited in that work is stored in separate files, limiting design creativity.

Multimedia software or Web-published portfolios allow students to make more choices regarding color, navigation, design, and layout. It is important that the first or second card, slide, or page contain an overview of the contents. If the first screen is an introductory one, be sure clear directions appear to steer the viewer to the Table of Contents. A word processing window can only give file names, but from this the viewer knows which file to open to view all of the contents within the electronic portfolio.

Most programs enable students to place navigation buttons on screens (as in Hyperstudio) or create links to supporting cards, slides, or pages within or outside of the portfolio. Decide the following when designing how to navigate through and view the portfolio:

- Use contrasting colors for the background and text. Preview choices to see if the text and images are easy to see. Background colors are good to use but often make the text hard to read.
- Whenever possible, insert images or graphics for clarity.
- Create and place navigational tools (buttons, links, etc.) so viewers can easily locate them. They should be apparent and consistent, preferably the same throughout the portfolio so viewers know where to click to proceed.
- Create a navigational tool on each card, slide, or page to let the viewer get back to the title cover or table of contents. However, viewers should not have to return to the beginning of the portfolio each time they want to view another area.
- Create a location for the teacher and other viewers to write comments to the author. This can be in the form of a separate *"Viewer Comment file"* where comments are added in a log or journal format. A card or slide could have a scrolling text box for comments,

or if the portfolio is Web-based, a link to the author's e-mail address can be provided through the software. If the software used to create the portfolio does not have this capability, create a separate word processing file on the same storage device (i.e., disk, hard drive, server) to enable viewers to write comments in a journal or log format.

Kid Pix Studio enables primary students to create electronic portfolios in a slide show format. The menu bar has large icons with sounds and pictures to accompany command choices, making it easy for younger children to work with minimal teacher direction. When introducing Kid Pix or any other software to young children, teachers should begin with whole-group instruction. Students should sit on the floor or bring chairs close to the large screen monitor to observe while the teacher demonstrate the following steps:

- Opening the program (for K–2 students creating an icon on the desktop works best)

Plugging in the Portfolio

SKILL LEVEL

- [ ] Entry
- [ ] Adoption
- [ ] Adaptation
- [x] Appropriation
- [ ] Invention

| File | Edit | Move | Tools | Objects | Colors | Options | Extras | Help |

Senior Project

I am pursuing an Aerospace Engineering degree at Cornell University. My senior research project focused on a major subset of issues dealing with transportation through the Earth's atmosphere and extraterrestrial travel.

Aerodynamics and control of flight and space vehicles, artificial satellites, propulsion systems, and all other aspects related to flight and space travel were discussed and defended to the Dean and his committee on May 18.

Complete data can be found on the school's web site: http://www.smithcountyHS.k12.us/seniorprojects

Written report available upon request.

SAMPLE 4.4 **This sample represents part of a presentation the student created in Microsoft PowerPoint for a Senior project. The student embedded the slide into the software used to create the electronic portfolio.**

- Opening a new or existing file
- Demonstrating tools (Using a variety of tools, the teacher begins to create a product similar to what is expected of the students.)
- Demonstrating how to save and close the program (Kindergarten students need extra help when saving work. It is recommended that the teacher and/or older student guide kindergartners one-on-one as they save their work. Sometimes the teacher or older student may have to just save it for them.

In a lab situation, a proven strategy for teaching students how to save their work is to stop the whole class and go through one step at a time. As the teacher models each step, let students follow along, monitoring and assisting them through the process.

Regardless of the software or storage device used, electronic portfolios teach students to become self-evaluators of their own literacy, growth, and learning. All portfolios will be different and unique to each student, and they are definitely worth the effort.

GAINING PROFICIENCY

Preparing for Productivity

The power of portfolios and the use of technology will only continue to increase in the coming years. Certain skills needed by students in the workforce, such as problem solving, finding and analyzing information, and collaborating with team members, cannot be measured on traditional multiple-choice tests. Viewing samples of student work displayed electronically in a portfolio is the best way to capture what a student knows, can do, and understands. To prepare all students to be productive workers in the twenty-first century, they must be given meaningful opportunities to demonstrate what they can do through performance tasks. Preparing all students to be productive workers today requires educators to provide meaningful opportunities for them to demonstrate what they know. The same skills workers are called upon to use on a daily basis in today's increasingly technologically based economy are some of the very same skills students cultivate and exercise when creating electronic portfolios.

Most performance tasks included in one's portfolio are assessed by a set of formal guidelines with several levels or dimensions laid out in a matrix. This matrix or rubric measures individual student work against specific criteria or evaluates the electronic portfolios as a whole. The true meaning of a score is determined by the design of the rubric. Deciding if work repre-

Plugging in the Portfolio

The sample portfolio in this chapter represents one created by a teacher at the Invention level of expertise. At this level, teachers are designing and implementing new models for technology use. New instructional patterns emerge from the use of technology. This sample is an individual portfolio with many cooperative group projects included. Students work in teams, focusing their learning though *vision categories* to deepen their understanding of different content areas. The classroom has moved from a teacher-centered environment to a student-centered one. Much collaboration takes place in this environment, and students rely upon themselves to solve problems and make critical decisions about their learning. This electronic portfolio was created in PowerPoint with many of the activities using other technologies: digital camera, scanner, word processing, charts, and word art.

sents quality and deep understanding depends on the precise wording, as it relates to quality and quantity, of each criterion level and if the rubric measures what should be learned as a result of performing the task. Several points should be considered when designing rubrics, including the following:

- Rubrics do not need to be task specific. A generalized approach is best, though they still need to address enough detail to create meaningful criteria.

- Limit the number of scoring dimensions to four or five. Breaking down the criteria into a greater number of categories hinders one's ability to score what really matters in the work.

- Be sure criteria is measurable, selecting descriptors carefully. Distinguish among the terms *few*, *many*, and *some;* these can hold different meanings for different people. Design the rubric to identify specific quantities with desired traits to support each. An example of measurable criteria that shows a specific quantity (in this case number of errors characteristic of this level) follows: *The electronic portfolio contains no more than three technical errors; however, the errors did not affect the overall navigation.*

- When possible, limit the rubric to four levels of performance. Fewer than four makes it difficult for the scorer to make a firm distinction. Too many levels make the rubric unmanageable and too cumbersome to use.

The analytical rubric can be used to assess electronic portfolios as a whole (see Figure 5.1).

Vision Categories

Adding technology to a portfolio system does not, by itself, change teaching and learning. According to Sheingold (1992), technology supports assessment when work is media accessible, portable, examinable, and widely distributed and when performances can be stored, retrieved, and viewed numerous times. It is the assessment component that drives a school-wide vision of reform. Real change in classrooms for *all* students occurs when electronic portfolios are organized with specific criteria.

Vision categories (or focuses within disciplines) are transferable ideas supported by concepts and/or skills. They are a means by which students can achieve deep understanding. Thus, vision categories can focus instructional delivery to make sure that concepts and skills are transferable across disciplines. Figure 5.2 contains a sample of six categories for developing meaningful electronic portfolios.

ANALYTICAL RUBIC

	ORGANIZATIONAL COMPONENTS	GRAMMAR & MECHANICS	NAVIGATION	CONTENT	REFLECTIONS
4—EXPERT PUBLISHER	Design and layout are logical, interesting, sequence is easy for audience to follow. Additions can easily be made as software options increase to include multiple forms of multimedia.	Although there may be a few insignificant errors within the work placed in the electronic portfolio, they do not detract from the presentation.	No assistance needed to navigate through portfolio. All work is easily accessible.	Student-centered, quality work denotes critical thinking skills. Students are provided numerous opportunities to collaborate. A variety of artifacts accurately demonstrate student understanding.	Understanding and purpose for learning the content is conveyed in the self-reflection (recorded verbally or in writing). All artifacts have a reflection attached. Student understands what the next step should be to extend learning. Student takes ownership in learning.
3—APPRENTICE PUBLISHER	Design and layout are logical; sequence is easy for audience to follow. Limited additions can be made if software options increase.	Although there may be a few errors within the whole work, they do not affect the portfolio as a whole.	Minimal assistance is needed to navigate through the portfolio. A few navigational errors may be present, but they do not substantially detract from viewing the portfolio.	Students are engaged in the work. It does provide opportunities for collaboration and revision but is traditional in nature. Work reflects low- level (rote) thinking skills. Student has some grasp of the content; however, it is difficult to determine if deep understanding exists due to limits of the activity.	Understanding of content is conveyed in self-reflection (written or verbal). Most artifacts have a reflection attached. Student conveys ownership of work.
2—NOVICE PUBLISHER	Design and layout are logical in parts; audience finds it difficult to observe learning that has occurred. Limited additions can be made if software options increase.	There are corrections noted within the work and in the portfolio itself affecting the display and meaning of the portfolio.	Assistance is needed to navigate through the portfolio. Work is disorganized and lacks flow, affecting the display and meaning of the portfolio.	Work is traditional in nature reflecting low-level thinking skills. No collaboration with peers. Limited variety of artifacts, difficult to determine what student understands.	Student explains work in a written or oral reflection, but it is unclear if student understands concept or skill. Few artifacts have a reflection attached. Student lacks ownership of his or her own work.
1—FURTHER MENTORING NEEDED	Design and layout are illogical throughout. Cannot determine if learning has occurred. No additions can be made if software options increase.	There may be many corrections noted within the artifacts, which detract from the meaning and usefulness of the portfolio.	Assistance is needed to navigate through the portfolio. The portfolio is disorganized to the extent that navigation is very difficult.	Teacher-centered, work is traditional, fact-based in nature, reflecting low-level thinking skills. Limited variety of artifacts does not demonstrate student understanding.	Cannot determine if an understanding was gained from task. No more than 10% of the entries have a student reflection attached (written or audio).
0	No work shown	No work shown	No work shown	No work shown	No work shown

Figure 5.1

The Steve A. portfolio in the samples that accompany this chapter uses vision categories to hone in on standards and goals. The design and layout used in the example support the use of vision categories and content standards. The viewer first sees an introductory screen with basic student information. Clicking on the student's picture leads the viewer to the navigational screen. This screen is designed as a table of contents. Student work is divided into six categories to correlate with a *vision statement* supported by each *content standard*. Including vision statements in the portfolio serves as a conceptual lens when examining how content standards fit into a real-world context. Clicking on a vision category enables the audience to view an assignment or reflection statement from a third screen.

Ways to Scale Up

If initial (pilot) electronic portfolios projects is begun with a defined group of students (i.e., one grade level, one class, or one school within a district)

VISION CATEGORIES FOR MEANINGFUL ELECTRONIC PORTFOLIOS

Following are a few examples of vision categories that can be used to meet standards that can be demonstrated via the electronic portfolio.

I. **Communication**—Supports each content standard: language arts, mathematics, foreign language, social studies, science, the arts, and health and wellness. One can successfully convey to others his or her knowledge, opinion, or perspective in different content areas to meet the academic standards.

II. **Acceptance of Self and Others**—A person's awareness of strengths and weaknesses allows them to reflect upon strategies for improvement. A natural outgrowth of this is developing an appreciation for cultural diversity and establishing goals that promote social responsibility and conflict resolution.

III. **Self-Reflection**—By examining past events, one can gain insight into how action and inaction shape culture and environment. Self-reflection allows a person to learn from his or her mistakes and to build on what they have done well in the past.

IV. **Critical Thinking and Problem Solving**—Deeper understanding of complex concepts is gained in all content areas through critical thinking and problem solving. Through this one is able to gather, analyze, synthesize, and evaluate information to make informed decisions.

V. **Values and Ethics**—Person demonstrates ethical behavior that includes honesty, justice, equity, cooperation, and respect for self and others regardless of cultural differences. Establishes a value system for decision making in the real world.

VI. **Conflict Resolution**—Decisions based on cooperation demonstrating understanding of various social, political, economic, or environmental points of view. Develop strategies to resolve differences to accomplish tasks.

Figure 5.2

and the goal is to expand to all students, a scale-up plan has to be developed. It is very helpful if teachers and students from the pilot project are involved in the scale-up planning sessions and mentor others as the project gets underway. The principal or administration team probably has the best idea of who would be the best person to act as a project leader and form a project committee to lead in the expansion. Good candidates for a leadership position should have the following characteristics:

- Are proficient with technology (They do not need to be functioning at the Invention level, but they should be able to perform basic troubleshooting tasks.)
- Are willing to share information with colleagues
- Are willing to revise their pedagogy and assessment practices
- Have a positive attitude
- Are well accepted by others and are accepting of others
- Are willing to put in extra time if necessary

Once a leader(s) is chosen, the next step is to schedule brainstorming sessions to determine the following in regard to the pilot project:

- What went well?
- What could have been done differently to improve any part of the project?
- Were there any critical technology issues that should have been addressed differently?
- Were there any assessment issues that should have been addressed differently?
- Were there management issues during any steps of the project that could have been done differently?
- Was there any resistance from students, teachers, parents, support teachers, or other staff members? If so, how could this be overcome in the scale-up plan?

The project leader or committee can decide the purpose and type of portfolio to begin the project, taking into consideration the current technology levels and use of alternative assessment of those who will work on the project. When the project is instituted school- or district-wide, teachers can decide what kind of electronic portfolio best fits their classroom. Only if the school institutes a graduation or high stakes portfolio system in which a student's work follows him or her from grade to grade does a committee need to define the criteria for a portfolio.

Leading by Example

Actual examples of electronic portfolios with some alteration or explanation make the best prototypes. If actual examples are not practical or available or do not represent best practice, the project leader or committee can create a prototype during introductory sessions for teachers, parents, community leaders, and students. The example does not need to be complete; it just needs to provide a visual sense of what a completed product might look like. The use of an example will also show the capabilities of current hardware and software and identify what technology should be acquired and in what order that should occur.

The scale-up plan should include the following steps:

- Survey current level of teachers' expertise in the use of technology and assessment practices (see Figure 5.3).
- Plan and schedule ongoing professional development sessions to train teachers based on survey.
- Survey quantity and capacity of current equipment.

Plugging in the Portfolio

SKILL LEVEL

- [ ] Entry
- [ ] Adoption
- [ ] Adaptation
- [ ] Appropriation
- [x] Invention

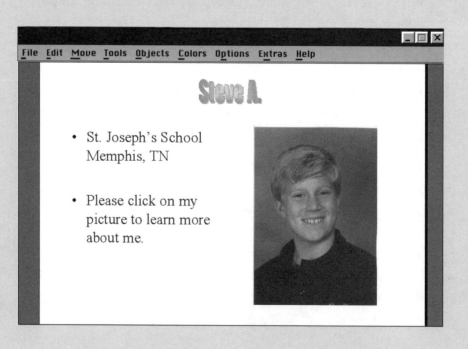

Steve A.

- St. Joseph's School Memphis, TN

- Please click on my picture to learn more about me.

SAMPLE 5.1 This sample represents an electronic portfolio that is student centered in that the student takes responsibility for its design, contents, and creation.

TEACHER SURVEY

	Entry	Intermediate	Proficient
SECTION ONE			
I. Desktop Management			
1. Turn computer off/on			
2. Re-boot when frozen			
3. Connect peripherals (printer, scanner etc.)			
4. Open and close windows			
5. Rearrange windows on desktop			
6. Create files and folders			
7. Use Apple Menu or My Computer			
8. Use the FIND command			
9. Initialize a disk			
10. Copy a file to and from disk			
11. Install and remove software			
12. Save and close a document			
13. Copy, Cut, and Paste from one document to another			
II. Computer Tools and Applications			
1. Word processing			
2. Use and/or create a database			
3. Use and/or create a spreadsheet			
4. Desktop publishing			
5. Charts/graphs			
6. Multimedia presentations			
7. Digital camera			
8. Scanner			
9. Image manipulating (Photoshop, etc.)			
10. Web page developing software			
11. Video camera			
12. Gather information from Internet			
13. Save an image off the Internet			

Figure 5.3

TEACHER SURVEY (CONTINUED)

SECTION TWO

	Entry	Intermediate	Proficient
III. Computer Use—Administrative Tasks			
1. Electronic grade book			
2. Communicate with parents/students via e-mail			
3. Electronic record keeping—attendance			
4. Intra-district communication			
IV. Teaching Practices—Has the use of technology changed instructional strategies?			
1. Less time spent lecturing whole group			
2. More time spent facilitating individual groups of students			
3. Comfortable with facilitating cooperative group activities			
4. Comfortable with students working independently			
5. Implement differentiated instruction			
6. Comfortable designing concept-based integrated units			
7. Comfortable managing classroom with multiple tasks			
8. Integrate technology as a tool to extend learning Give example:			
9. Use content standards/state achievement objectives as a guide to plan units and activities			
10. Comfortable identifying appropriate technology to use			
11. Adjust to different levels of students' technology expertise			
12. Assess student products with various instruments Give examples:			
13. Design holistic and analytical rubrics			

Figure 5.3 (continued)

Plugging in the Portfolio

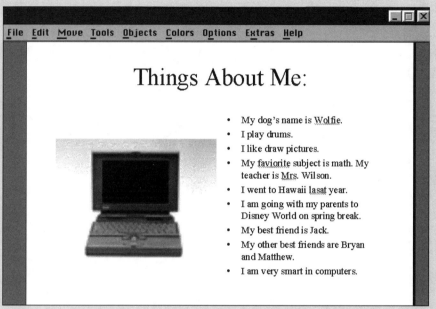

SAMPLE 5.2 Depending upon their interests, students add clip art to support the text. The viewer presses the down key on the keyboard to proceed to the next slide.

SAMPLE 5.3 The teacher sets up a template for students to complete and import into their portfolio material to help them focus on goals and vision categories. The viewer can click on "Steve A." to advance to the next slide.

Plugging in the Portfolio

SKILL LEVEL

- [ ] Entry
- [ ] Adoption
- [ ] Adaptation
- [ ] Appropriation
- [x] Invention

File   Edit   Move   Tools   Objects   Colors   Options   Extras   Help

Portfolio Visions

My portfolio has several Vision Categories to help me reach my goals for my work. The work chosen and added to this portfolio represent standards I have met within my units of study. Units in the electronic portfolio are listed below:

I. Unit: "Off To War We Go" (Social Studies Focus)

  Concept: Conflict

  Activity: Letter to Gulf War Soldiers

  Vision for Literature/Media

Through Communication:

•Human populations share common needs to survive.

•Communities share similar characteristics in complex societies.

II. Unit: "World Beginnings" (Religion)

Concepts: Individual Development

Activity: Essay "Creation Story"

Vision for Literature/Media

Through Communication:

•Cultural Beliefs and perspectives shape our faith.

SAMPLE 5.4 Students are given a template to complete to help focus their learning and think about why certain work is chosen for inclusion, which builds critical and reflective thinking skills.

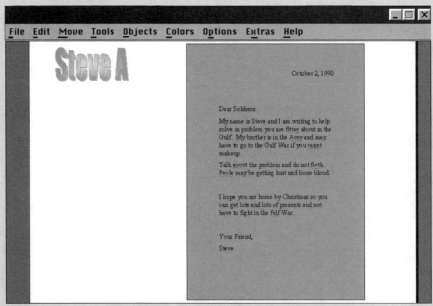

File   Edit   Move   Tools   Objects   Colors   Options   Extras   Help

Steve A

October 2, 1990

Dear Soldiers:

My name is Steve and I am writing to help solve in problem you are fiting about in the Gulf. My brother is in the Army and may have to go to the Gulf War if you cannt makeup.

Talk avout the problem and do not firth. People may be getting hurt and loose blood.

I hope you are home by Christmas so you can get lots and lots of presents and not have to fight in the fulf War.

Your Friend,

Steve

SAMPLE 5.5 The letter was created using word processing. The image was scanned in but could also have been imported from a file made with a digital camera.

Plugging in the Portfolio

SKILL LEVEL

☐ Entry
☐ Adoption
☐ Adaptation
☐ Appropriation
☒ Invention

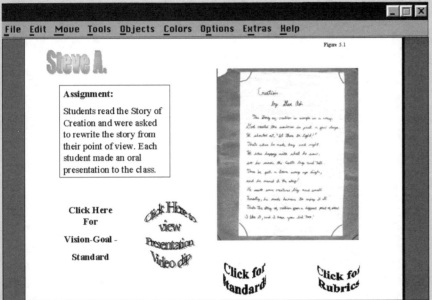

SAMPLE 5.6 This template is used to organize student work placed in the electronic portfolio. The original student work is captured with a digital camera and imported into the portfolio. Buttons can be placed on the screen leading the viewer to information on the Internet, such as district standards or a video clip for rubrics.

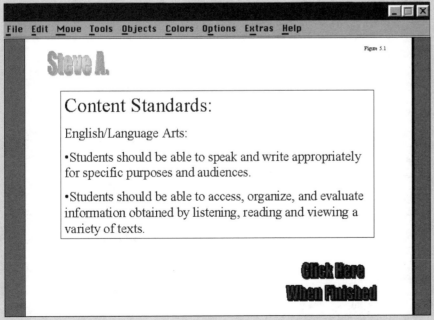

SAMPLE 5.7 A template is included for the student to identify the content standard a particular assignment is designed to meet.

Plugging in the Portfolio

SKILL LEVEL

- ☐ Entry
- ☐ Adoption
- ☐ Adaptation
- ☐ Appropriation
- ☒ Invention

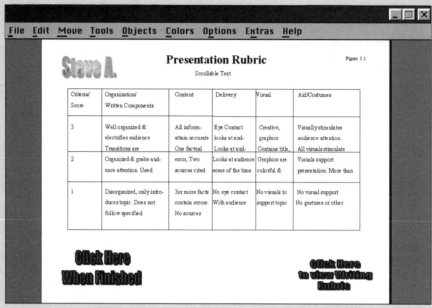

File Edit Move Tools Objects Colors Options Extras Help

Steve A.   Presentation Rubric   Figure 5.1
Scrollable Text

Criteria/ Score	Organization/ Written Components	Content	Delivery	Visual	Aid/Costumes
3	Well organized & electrifies audience Transitions are	All inform- attain accurate One factual	Eye Contact looks at aud- Looks at aud-	Creative, graphics Contains title,	Visually stimulates audience attention. All visuals stimulate
2	Organized & grabs aud- ence attention. Used	error, Two sources cited	Looks at audience some of the time	Graphics are colorful &	Visuals support presentation. More than
1	Disorganized, only intro- duces topic. Does not follow specified	3 or more facts contain errors. No sources	No eye contact With audience	No visuals to support topic	No visual support No gestures or other

Click Here When Finished

Click Here to view Writing Rubric

SAMPLE 5.8 This is a screen capture of a rubric table the student found on the district's Web site. Text is scrolling; therefore, all words do not appear here. The student can place the rubric in the electronic portfolio to show the criteria used for assessment.

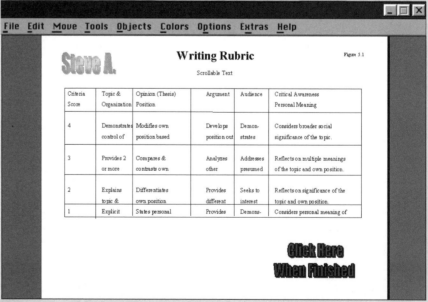

File Edit Move Tools Objects Colors Options Extras Help

Steve A.   Writing Rubric   Figure 5.1
Scrollable Text

Criteria Score	Topic & Organization	Opinion (Thesis) Position	Argument	Audience	Critical Awareness Personal Meaning
4	Demonstrates control of	Modifies own position based	Develops position out	Demon- strates	Considers broader social significance of the topic.
3	Provides 2 or more	Compares & contrasts own	Analyzes other	Addresses presumed	Reflects on multiple meanings of the topic and own position.
2	Explains topic &	Differentiates own position	Provides different	Seeks to interest	Reflects on significance of the topic and own position.
1	Explicit	States personal	Provides	Demons-	Considers personal meaning of

Click Here When Finished

SAMPLE 5.9 Teachers have access to a district database of rubrics to use for assessing student work. The student was able to capture a screen shot of the rubric table (text is scrolling; there- fore, all words do not appear in the screen shot) and place it in the electronic portfolio to show the criteria used for assessment. Since the assignment in Sample 5.6 used two rubrics in the assessment, this second one was added to the portfolio.

Plugging in the Portfolio

With the emphasis on concept-based integrated units and the changes in the way textbooks are used for instruction, teachers are feeling overwhelmed and frustrated as they begin planning activities for their classroom. It is difficult to plan meaningful learning experiences for students (instructional tasks, lessons, units, rubrics, etc., ensuring they are linked to standards, goals, visions, skills, higher-level thinking, intelligences). Don't forget about the students that read below or above grade level! One gets tired just thinking about it! Without instructionally appropriate strategies, students cannot produce quality work for inclusion in an electronic portfolio. There are just not enough hours in a day.

Administrators and teachers are looking for ways to make creating the "what" easier so they can concentrate on the "how." As technology has made the creation of portfolios more manageable, it was determined it could have the same effect on teacher planning. Recognizing these challenges, IBM and educators in several districts and states have collaborated through Reinventing Education Partnerships to create innovative tools that go hand-in-hand with the creation of electronic portfolios.

Using Wired for Learning technology, these partnerships work with resources from the Internet to design exemplary instructional activities. To ensure students will produce quality work, each activity is subject to a rigorous, online jurying process before other educators can share it for use. Teachers assess student work online with an assessment scoring tool as shown in Sample 5.10 to produce a report or student profile that can be added to the electronic portfolio. Teachers can easily identify academic strengths and challenges for each student, as well as those from the entire class. The tool thus helps teachers improve instruction meeting the needs of all students.

Web site: **http://www.ibm.com/ibm/ibmgives/grant/education/programs/reinventing/wfl.html**

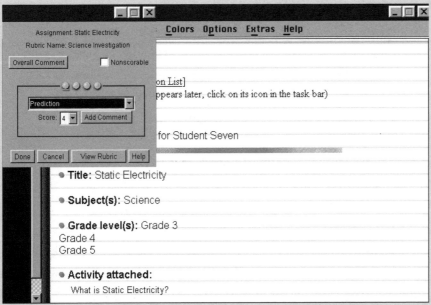

SAMPLE 5.10 Once students complete their product, teachers can score the work with an online assessment scoring tool. The criteria from the rubric or scoring guide appear in the scoring window. Student reports are created from the data collected for use in conferences and to document student progress.

- Develop a plan for how and where students and teachers can access necessary equipment.
- Develop a plan for acquiring technology.
- Develop a mentoring plan for teachers to include regular sharing sessions.
- Develop a monitoring and evaluation plan for the project.

Change is hard for everyone, and it is especially difficult when it involves the entire faculty. The key to encouraging an entire faculty to incorporate electronic portfolios into classroom practice is to provide continuous professional development. If funding can be secured to pay teachers a stipend for such training, they will be less likely to resist devoting time outside of the normal school day to the acquisition of the necessary skills. In addition, it is important to develop a schedule of when different technology-based activities will begin in the classroom. Coordinate training to give teachers enough time to practice what they have learned but no so much time that they *lose it because they didn't use it.*

If a teacher has always conducted teacher-directed activities, arranged students in rows to work individually, and maintained silence, then using an electronic portfolio system is going to be a drastic change. However, a well thought out plan with numerous opportunities for continuous support for teachers and students will help create a successful project. All students deserve the opportunity to tell their story. Technology can overcome obstacles, making dreams come true for all students.

Presenting electronic portfolios as a new medium for displaying and/or assessing student work may be a new experience for many. The project leader and/or committee needs to be aware of the types of comments and questions that will be asked by the audience. Be aware the audience may be fascinated by the technology and not fully understand the portfolio's full potential and how it can be used to raise student achievement. The key to changing a school's culture is communicating the importance of why students must demonstrate what they know, can do, and understand through authentic work. It is more important to show an evolving process and the steps a student took to complete the task than to just take a "snapshot" of the final product. Explaining these points to students, parents, teachers, and the community will help them embrace electronic portfolios as a necessary tool for demonstrating student abilities. The key is to promote continual communication among all involved.

As more and more districts require students to create portfolios to "show what they know" and access to technology becomes commonplace within all schools, electronic portfolios will become the standard means to communicate progress throughout one's K-12 education.

MANAGING A RESTRUCTURED CLASSROOM TO INCLUDE ELECTRONIC PORTFOLIOS

So Much to Do, So Little Time

One of the most important measures of a successful teacher is how well he or she manages the learning environment. Classroom management is essential if learning is to take place and students are to succeed. Learning occurs in a structured environment with predetermined procedures and routines. When teachers establish behavior standards on the first day of school, the chances of students experiencing success increases because students know at the beginning what is expected of them.

Some educators think that covering chapters in a textbook, answering questions, then giving a multiple-choice test "covering" the material is what teaching is all about. In such a setting, computers are used in classrooms only as a reward for finishing an assignment ahead of one's classmates. According to Wong and Wong, "it is the responsibility of the teacher to manage a class to see that a task-oriented and predictable environment has been established" (1998, 88). Without classroom procedures, students waste time between tasks and chaos arises.

Effective teachers use a variety of instructional strategies to promote learning for all students. For many activities, including electronic portfolio work, using well-managed cooperative groups provides an opportunity for students to help each other throughout the learning process. Researchers such as Slavin (1995) have studied the benefits of

Plugging in the Portfolio

The portfolio sample offered in this chapter is an individual student's K–12 portfolio. Such a portfolio may result from a district establishing standards for implementation of electronic portfolios that would follow a student from kindergarten through twelfth grade. Software applications, portfolio design, and other technologies available are those that teachers at the Adaptation level could effectively utilize. The level of proficiency chosen should reflect the level at which most teachers in that district are working. Additional entries through the years can vary from teacher centered to student centered depending on how comfortable the respective teachers are with technology.

cooperative group learning for both high and low achievers. He found that working in cooperative groups is beneficial for both groups only if teachers follow the following guidelines:

- Reward all students regardless of ability for successfully working together within the team.
- Hold each student accountable for learning the content and skills.
- Reward all students if they put forth the effort

As classrooms move from teacher-directed to student-centered learning environments that include the integration of technology, new management strategies are required. When classroom computers and other technologies are introduced, teachers must be prepared to establish procedural steps for students to follow to make efficient use of all the equipment. Effective teachers spend the first two weeks of the school year teaching students classroom procedures and daily routines including the use of technology as a tool for their learning. Teachers should establish procedures for students to follow for the following activities prior to introducing electronic portfolios in the classroom:

- Listening/responding to questions
- Indicating understanding
- Working cooperatively
- Changing groups
- Collecting work for a portfolio
- Conferencing with peers/teacher
- Getting materials for activity
- Moving about the classroom
- Going to the library-media center/computer lab
- Completing assignments early
- Handling technology
- Scheduling use of technology

Teacher Roles

Teachers play a variety of roles in a student-centered classroom when electronic portfolios are used. These roles are often dependent upon the age level of the students. With primary students, create group or whole class portfolios that allow each student to participate in the process. Older students can create their portfolios with fewer interactions with the teacher and more collaboration among peers. Peer learning is an effective strategy when the following steps in planning and implementation are followed:

Teacher Planning

- Identify the topic or content to be covered.
- Prepare instructional materials, resources, and schedule technology use.
- Assign peer partners.
- Teach students strategies to give constructive feedback, encourage verbalization, and encourage partner to stay on task.

Implementation

- Conduct whole group presentation to set the stage or give directions or new content.
- Break into peer partner pairs.
- Monitor progress moving among groups.
- Monitor and evaluate peer pairs.

Establishing procedures prior to beginning electronic portfolios, practicing appropriate procedures, and setting high expectations for students to work independently in pairs will result in a well-managed class-

Plugging in the Portfolio

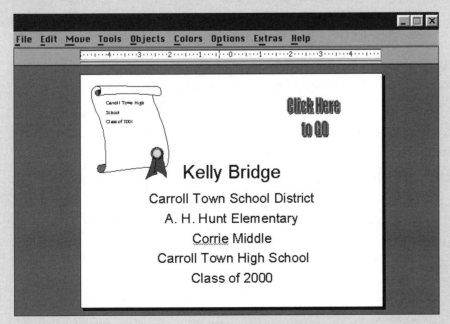

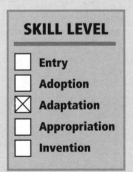

SKILL LEVEL

- ☐ Entry
- ☐ Adoption
- ☒ Adaptation
- ☐ Appropriation
- ☐ Invention

SAMPLE 6.1 This sample represents work collected throughout a student's K–12 career. A district level committee determined the criteria for the contents of the portfolio. It also decided that portfolios would be created with tools familiar to teachers who are at the Adaptation stage since most teachers have not progressed past that technical level.

room. Even as students continue to assume more responsibility for constructing their electronic portfolios they may still request guidance from the teacher. When this is the case the teacher should help them re-examine their objectives and ask guiding questions to lead them toward *their* goal.

Reorganizing the Schedule

In student-centered classrooms it is not uncommon to see several different activities going on at the same time. The creation of electronic portfolios just becomes another part of the project. Time management is critical and presents new challenges because all students need access to resources and equipment. Scheduling electronic portfolio work will depend on the amount of equipment available.

If one computer is in the classroom, students can rotate individually or in small groups. Teachers should be sure students have everything they need to use their computer time productively. Students who are working on noncomputer-based tasks can be working on selecting, reflecting, design, or layout of their portfolio in preparation for their computer-based activities (see Figure 6.1).

If students have access to several computers at the same time it allows a great deal of flexibility in scheduling, but it still requires planning on the part of the teacher. Because electronic portfolio development occurs over time, designate one computer (or group of computers) to be used for electronic portfolios. Locate peripheral equipment, such as the scanner or video camera, and check to see if it is only available at certain times for student use. Different software applications can be installed on certain specified computers. Schedule students on a rotating basis so all have access to the applications needed to complete their tasks (see Figure 6.2).

If some of the portfolio work will be done in a computer lab setting, structure the activities so that all students are to a point where they can do the computer-based tasks. If all of the lab computers have the appropriate software used for electronic portfolio development, then all students can work on the same activity. If computers in the lab have different software applications, extra planning will need to be done by the teacher to ensure students have a computer-based task to match the available software or that their task matches the type of equipment they are scheduled to use (see Figure 6.3).

COMPUTER SCHEDULE OPTION I

	Monday	Tuesday	Wednesday	Thursday	Friday
Computer 1	Group 1	Group 3	Group 5	Group 2	Group 4
Computer 2	Group 1	Group 3	Group 5	Group 2	Group 4
Computer 3	Group 2	Group 4	Group 1	Group 3	Group 5
Computer 4	Group 2	Group 4	Group 1	Group 3	Group 5

Figure 6.1

COMPUTER SCHEDULE OPTION II

	Monday	Tuesday	Wednesday	Thursday	Friday
2 classroom computers—Internet research	Group 1	Group 2	Group 3	Group 4	Group 5
2 classroom computers—electronic portfolios	Group 2	Group 3	Group 4	Group 5	Group 1
2 library/media center computers (scanning, word processing)	Group 3	Group 4	Group 5	Group 1	Group 2
Task not using technology	Group 4	Group 5	Group 1	Group 2	Group 3
Task not using technology	Group 5	Group 1	Group 2	Group 3	Group 4

Figure 6.2

COMPUTER SCHEDULE OPTION III

	Monday	Tuesday	Wednesday	Thursday	Friday
Classroom computer	Group 1	Group 2	Group 3	Group 4	Group 5
Media center/ library or computer lab	Group 2	Group 3	Group 4	Group 5	Group 1
Task not using technology	Group 3	Group 4	Group 5	Group 1	Group 2
Task not using technology	Group 4	Group 5	Group 1	Group 2	Group 3
Selection/ Reflection meeting with teacher	Group 5	Group 1	Group 2	Group 3	Group 4

Figure 6.3

Peripheral equipment, whether it resides in the classroom or is borrowed from an outside location, must be scheduled so all students have access to it. Sequencing of all supporting tasks leading up to what is done with the technology must be completed in order to make efficient use of time scheduled for technology-based tasks. The best practice is for teachers to require each student to discuss with the group's project manager the task they plan to work on during their scheduled time on the computer, scanner, digital camera, or other equipment. (See the following discussion for student roles within groups.) This transfers the decision-making processes to students, increasing their ownership of the project. If a student does not have the necessary tasks completed prior to using the equipment, the project manager can choose to let another student use the time, which is noted to the teacher when the project manager fills out the Daily Work Plan (see Figure 6.4).

When primary students are assuming some responsibility for electronic portfolio development, it helps to pair them with upper grade students to oversee the project management tasks. Younger children need extra help opening files, saving to disks, and organizing materials. Setting up a mentoring system is an excellent learning experience for all ages.

DAILY WORK PLAN

Group Name: _____ Date: _____

Assigned Center for Today:_____

Member Name:_____

Prepared for Task (Y/ N) Completion Goal for the day:_____

_____

_____

Comments: (Required for any group member not prepared for center activity when scheduled. What will that student do during project time)?_____

_____

_____

Project Manager Signature _____

Figure 6.4

Helpful Strategies for Managing Equipment Use: Primary Grades

In the primary grades, a well-managed electronic portfolio project begins with a teacher designed project plan. See Figure 6.5 for an example of a teacher-created project plan. Organizing the project to ensure the skills primary age students should know creates a meaningful collection of work. Each student can create a portion of the portfolio using a jigsaw strategy for cooperative groups. This will eliminate repetitious examples of student work and create variety. Each group of students is responsible for different information. Asking older students to act as mentors to help with some of the multistep tasks will help manage students engaged in differentiated instructional tasks.

Once information has been gathered, young students, with the help of their older mentors, can design a storyboard to lay out how the different steps will be arranged in the electronic portfolio. When students are ready to use the technology to create the electronic display, their information can be assembled in the computer lab or in the classroom with the assistance of older student mentors or the teacher. Teachers reduce confusion and eliminate unnecessary questions if they take the time to post clear step-by-step directions for creating the portfolio, using equipment, and describing each group member's role. If students are absent or need extra time, one solution is to have an extra computer set up in the library, lab, or other central location to use when extra time is available.

Helpful Strategies for Managing Equipment Use: Upper Middle Grades

An effective strategy for a well-managed upper elementary classroom begins when the stage is set at the beginning of the day. When students arrive for the day they check the equipment schedule to determine which

Plugging in the Portfolio

SKILL LEVEL

☐ **Entry**

☐ **Adoption**

☐ **Adaptation**

☐ **Appropriation**

☒ **Invention**

Science Project Plan 2nd grade

Research and identify annual and perennial flowers

Plan, calculate & purchase tools, materials and seeds to plant both types of flowers

Plan, calculate & Purchase tools, materials and seedlings to evergreens

Research Wildflowers to determine which type to plant

Group Four

Plan, calculate & Purchase tools, materials and seeds to plant wildflowers

Third Group

Biodiversity Environmental Study on the School's Nature Trail

First Group

Research evergreens to determine which type to plant

Second Group

Research deciduous trees to determine which type to plant

Plan, calculate & Purchase tools, materials and seedlings to plant deciduous trees

SAMPLE 6.2
In the primary grades, a well-managed project begins with a teacher-designed project plan. In this sample, the teacher created the plan in Inspiration software and placed the digital file in the whole class electronic portfolio to show what topic each group studied.

groups have use of the computers during project time. Looking at working folders located in the project bin, students can organize their work to proceed with the next task in the creation of their portfolios. All references and supplies need to be organized before students are ready work on the computer. Each student shows the project manager his or her layout sheet or other resources for sign-off on the Daily Work Plan.

Helpful Strategies for Managing Equipment Use: Middle and High School

Implementing a middle/high school electronic portfolio project takes real effort. The entire staff must coordinate how students can have access to technology, organize, and assemble work from different classes and subject areas and have the necessary time to work together and reflect meaningfully upon their work. An advisory board consisting of the administration and teachers from each department can organize how these tasks will be accomplished. Investigating different scheduling solutions may allow students to increase collaborative time for their project and not severely affect

Plugging in the Portfolio

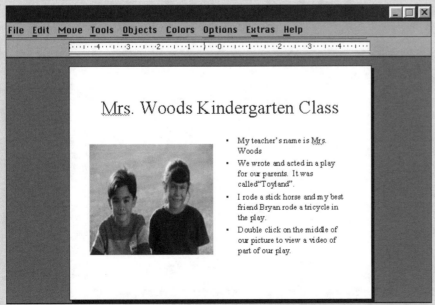

Mrs. Woods Kindergarten Class

- My teacher's name is Mrs. Woods
- We wrote and acted in a play for our parents. It was called "Toyland".
- I rode a stick horse and my best friend Bryan rode a tricycle in the play.
- Double click on the middle of our picture to view a video of part of our play.

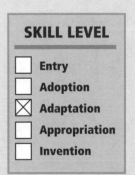

SKILL LEVEL

☐ Entry
☐ Adoption
☒ Adaptation
☐ Appropriation
☐ Invention

SAMPLE 6.3 The student's kindergarten class created introduction slides for an "Open House," after which, the teacher placed each student's information slide in their K–12 electronic portfolio. Student pictures were taken with digital cameras.

quality class time. Extending the homeroom period once or twice a week, extending the school day by eight minutes in the morning and six minutes in the afternoon, and creating an additional short project period after the lunch break are some strategies several middle and high schools have used. Changing to a block-scheduling scheme may also facilitate cooperative work and the meaningful use of technology.

Grouping students and assigning roles within each group should be planned ahead of time by the teacher to reduce confusion during activities. Make it a practice to rotate roles and members within groups according to the assignment, enabling all students to experience different roles. Students can be grouped in the following ways:

- Heterogeneously
- Homogeneously
- By dominant Intelligence (Bodily/Kinisthetic, Verbal/Linguistic, etc.)
- Randomly grouped
- To include a technology expert

Plugging in the Portfolio

SKILL LEVEL

☐ Entry
☐ Adoption
☒ Adaptation
☐ Appropriation
☐ Invention

3rd Grade Writing Assessment

SAMPLE 6.4 Often district committees require all K–12 portfolios to contain a successful Writing Assessment from the 3rd grade. The essay can be scanned into a digital format and imported into Microsoft PowerPoint for addition into the electronic portfolio as can assessments from later grades.

Figure 6.5 lists possible jobs for group members.

GROUP MEMBER JOBS

- Reporter – This person reports status and progress to the class and/or teacher.
- Resource Manager – This person is responsible for gathering materials needed prior to starting the activity, checking out books from the library or reference center and returning materials at the end of the period.
- Documentation Manager/Recorder – This person records data, takes notes or creates a concept map during brainstorming sessions.
- Technician – This person operates equipment, ensures equipment is correctly connected, returns equipment after use, and reports to the teacher if equipment is not working correctly.
- Project Manager – This person keeps all group members on task and on time and schedules time for use of equipment, resources, teacher, or peer conferences.

Figure 6.5

Plugging in the Portfolio

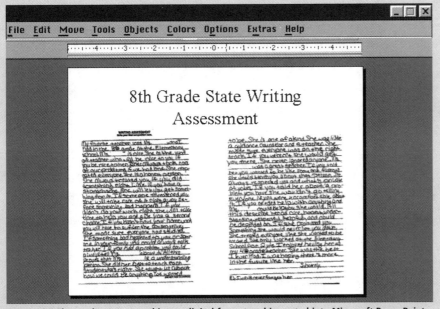

SAMPLE 6.5 The work was scanned into a digital format and imported into Microsoft PowerPoint.

Classroom Layout

Review Figure 6.6 for a possible classroom layout. Many disruptions can be avoided if the room is arranged in a way that students know the teacher is close by and that he or she will make the effort to intervene when necessary. Furniture should be arranged so that traffic patterns facilitate teacher and student movement. The teacher should be able to approach students easily to monitor their progress or offer help during the activity. Some students, especially at the secondary level, will slouch down in their seats with feet sticking into the walkway. Avoid narrow walkways between desks.

Educators must work around existing doors, outlets, bookcases, and any other fixed structures. Tech-savvy computer floor planners were not around when many schools were built. Because there are no outlets in the center of the room, equipment must be placed around the perimeter. Placing computers around the room instead of grouping them together in one area of the room gives students collaborative workspace. Some computer workstations are dedicated to specific functions (i.e., electronic portfolios, scanning, Internet research, embedding the electronic portfolios into the curriculum).

Arranging student seating in clusters (tables or groups of desks) promotes collaboration. Many educators have the misconception that teacher-directed instruction does not take place in student-centered classrooms. This is far from the truth. At certain times, all classrooms must have teacher-centered activities to deliver new knowledge. Notice how no student is seated with his or her back to the whole group instructional area. By turning their heads slightly, all students can see the overhead projector or teacher station for teacher-directed activities when necessary.

Ideally, electrical outlets should be placed in close proximity to student work areas so computers are accessible at all times. For example, if students are working on a science experiment at their cluster table and a computer is placed in the work area, results could be discussed with another class performing the same activity via the Internet. Data could then be placed in a collection folder on the computer for later inclusion into the electronic portfolio selection process. This saves time, promotes collaboration, and eliminates the extra step of transferring data from one format to another.

The Wireless Classroom

When laptops are used, there is little need to worry about the availability and placement of electric outlets. Laptops can be used on desks or tables

CLASSROOM LAYOUT

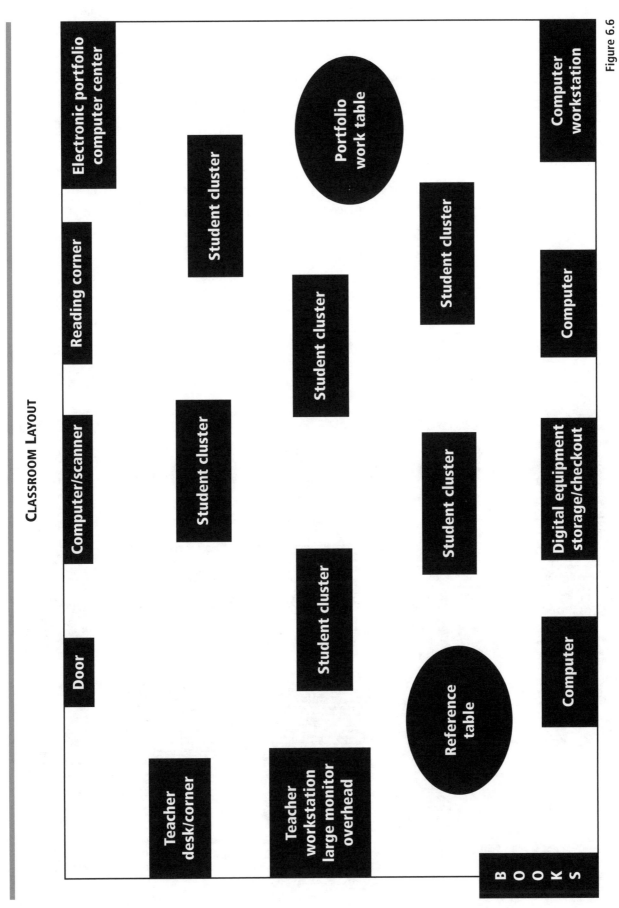

Figure 6.6

or taken to the library or outside to conduct research. Wireless technology works especially well for collaborative peer reflection. Students can sit together and discuss a piece of work while writing their reflections on their laptop computers. Many laptops even have wireless downloading capabilities to transfer information from the laptop to a desktop computer.

If other pieces of equipment, such as calculators, digital cameras, or electronic probes, are stored in the classroom, they should be locked in a cabinet with easy access for the teacher and/or the equipment manager to ensure they are checked out and returned in an orderly manner. When a computer is placed with a group's work area, be it a laptop or desktop, it is more likely that it will be used as a tool to extend learning instead of as an add on activity to "regular" curriculum activities.

Integrating electronic portfolios into curriculum, instruction, and assessment requires the teacher to carefully consider the use of time and space. Teachers should set goals for each work session. After an initial session, teachers need to require students to place one or two entries into the portfolio during each session at the computer.

Plugging in the Portfolio

SKILL LEVEL

- [ ] Entry
- [ ] Adoption
- [x] Adaptation
- [ ] Appropriation
- [ ] Invention

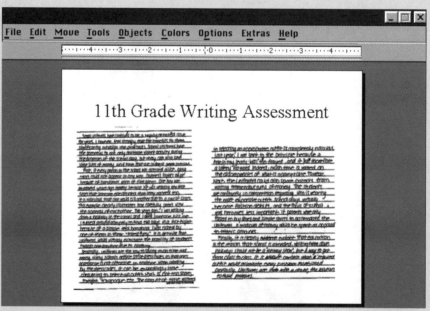

SAMPLE 6.6 The work was scanned into a digital format and placed in the electronic portfolio. Other entries are included to demonstrate competencies in all academic areas at various grade levels.

Since most schools have more students than computers, one strategy for managing activities is to pair students during project time. While each student is responsible for his or her own portfolio, they can work together, concentrating on one student's portfolio during one session and on the other's the next time.

Whatever managing strategies work best, remember to plan each step in the process and set up check points along the way to guide students to use their time and the equipment efficiently.

COMMUNICATING ACHIEVEMENT

Using Electronic Portfolios to Strengthen Authentic Assessment

Multiple-choice tests have been commonplace in most schools for more than a century. After the teacher graded each test, it was returned and a score was recorded in the grade book. The class moved onto the next chapter, rarely revisiting or discussing the previous chapter any further. In such a scenario, no feedback is given to the teacher to determine if all students understand the information. This makes it difficult for the teacher to know if any adjustments should be made in the next lesson. No meaningful discussion takes place between the teacher and student.

As discussed in chapter 3, a portfolio classroom looks at assessment in a new light. Students take ownership of their learning as they carefully evaluate their performances in the collection, selection, reflection, and projection processes. Revisions are encouraged and, in fact, recommended prior to receiving a grade. It is not uncommon for a teacher to return work to a student with suggestions on how to improve it before turning it in for a final grade.

By the time parent conferences are scheduled in a traditional assessment environment, the best the teacher can share with parents are numbers that represent a percentage of correct answers their child received on a particular assignment or group of assignments. The percentage of correct answers does not tell a story about how

Plugging in the Portfolio

The sample portfolio in this chapter is representative of one created with special needs students and illustrates work at the Adoption level. Teachers at this level can successfully use technology at a basic level and regularly assign "buddy" students to help their special needs students create class or individual electronic portfolios. With a basic knowledge of software application, teachers are able to plan structured work sessions. Many activities are teacher-centered, and teachers take an active role in helping the "buddies" know how to best help the special needs students. Because classroom management strategies vary among special education classes, a group portfolio is illustrated here. Teachers should try to pair the same "buddies" each time. The "buddy" link is critical for a successful experience.

the student felt that day, if a sad event happened on the way to school, or even if the concept was understood. It only tells what a student was able, or unable, to do at a particular point in time.

Portfolio classrooms view conferences as the vital link in a student's K-12 education. According to Farr and Tone (1994), the most important part of portfolio assessment is the use of regularly scheduled conferences between teachers and students, teachers and parents, students and parents, and among all three partners. Students should also reflect upon each other's work in peer discussions.

As discussed throughout this book, students' reflections are critical in the assessment process. Thinking about one's own learning enables students to engage in meaningful conversations about the knowledge that was gained and how this knowledge can apply to authentic situations. Reflecting upon what was just learned helps students determine the next step(s) they should take to deepen their understanding about the concept or skill.

Involving students in the planning, scheduling, dialogue, and evaluation of conferences sends an important message that what students have

Plugging in the Portfolio

SKILL LEVEL

☐ Entry
☒ Adoption
☐ Adaptation
☐ Appropriation
☐ Invention

SAMPLE 7.1 This sample represents a group portfolio created with special needs students.

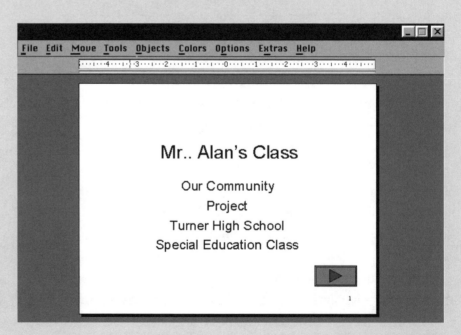

File   Edit   Move   Tools   Objects   Colors   Options   Extras   Help

Mr.. Alan's Class

Our Community
Project
Turner High School
Special Education Class

1

to say about their learning—the products and performances they create and the self-reflections they write—is important. Engaging students in these steps increases self-esteem and strengthens their commitment to an electronic portfolio project and consequently their own learning.

Many reasons exist to schedule conferences to discuss student work. Defining the purpose, setting goals, and organizing an agenda are the first steps to ensure a successful experience for all parties involved. Following are the different purposes for planning portfolio conferences:

- Teacher/Student—Review progress and set goals for future learning. Celebrate completion of the culminating events in a project, unit, or course. Increase students' understanding of what is expected from them prior to beginning a new task.

- Student/Student (same grade level)—Broaden understanding about a product or performance through peer discussions. Peers determine if the work represents "quality" and if any changes could occur to improve the work. Promote teamwork and the ability to communicate effectively.

Plugging in the Portfolio

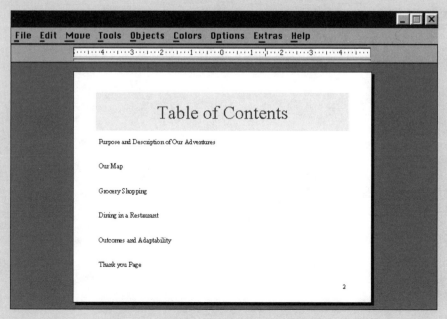

SKILL LEVEL

- ☐ Entry
- ☒ Adoption
- ☐ Adaptation
- ☐ Appropriation
- ☐ Invention

SAMPLE 7.2 This sample is created with a teacher who is at least at the Adoption level. The portfolio is teacher-centered with assigned "buddies" to help students with the actual construction and layout.

Table of Contents

Purpose and Description of Our Adventures

Our Map

Grocery Shopping

Dining in a Restaurant

Outcomes and Adaptability

Thank you Page

2

- Student/Student (multiage groups)—Increase mentoring oppotunities among students across grade levels.
- Student/Teacher/Parent—Increase students' involvement in their own learning. Help students gain an understanding for future improvement. Convey student progress to parents. Set future learning goals. Virtual conferences can be taken to a new level, especially with the increased mobility in today's society.
- Administration/Teacher—The use of electronic portfolios as evidence of "best practices and strategies" used by a teacher is an effective way to demonstrate methods used to teach to high standards. If a local area network exists in the school, the teacher can place lesson and unit planning on a local school server for easy viewing. Electronic portfolios that store the products students produce can then be viewed from the server without delay. The technology provides flexibility for principals to look inside the classroom through electronic portfolios and assess the quality of work provided by teachers.

Plugging in the Portfolio

SKILL LEVEL

- [ ] Entry
- [x] Adoption
- [ ] Adaptation
- [ ] Appropriation
- [ ] Invention

SAMPLE 7.3 Using Hyperstudio or Microsoft PowerPoint, each team is responsible for designing and constructing one slide for the portfolio. In this sample, students captured a map from the Internet, saved the image on a disk, and imported it into their electronic portfolio with the help of a "buddy."

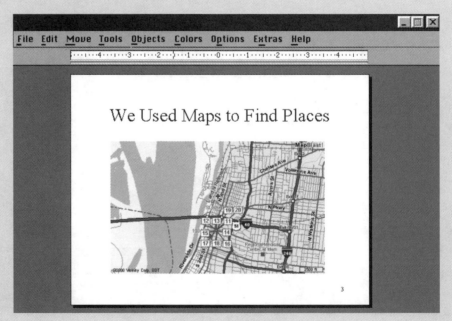

- School/Community—Demonstrate accountability through the use of electronic portfolios to show the skills students are gaining as they prepare for the technological rigor of the work place.

As teachers and students choose the purpose and audience for conferences, a schedule must be developed to accommodate all students. These time frames are suggested and should be adjusted to meet the needs of a particular grade level or number of students:

- Teacher/Student—Twice per grading period or at the beginning/end of a new project.
- Student/Teacher/Parent—End of each grading period, if requested, otherwise one per semester.
- Student/Student (multiage groups)—Twice per month.
- Administration/Teacher—One per school year, unless contract specifies otherwise.
- School/Community—One per school year.

The next steps include determining who will be involved in the portfolio conference and the purpose for the meeting, choosing the loca-

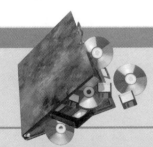

Plugging in the Portfolio

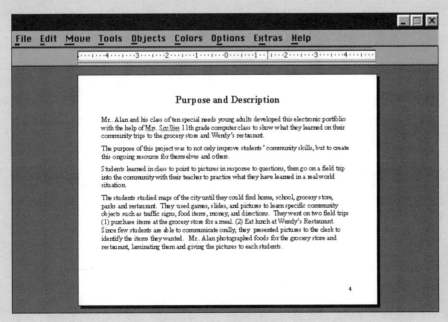

Purpose and Description

Mr. Alan and his class of ten special needs young adults developed this electronic portfolio with the help of Mrs. Scullies 11th grade computer class to show what they learned on their community trips to the grocery store and Wendy's restaurant.

The purpose of this project was to not only improve students' community skills, but to create this ongoing resource for themselves and others.

Students learned in class to point to pictures in response to questions, then go on a field trip into the community with their teacher to practice what they have learned in a real world situation.

The students studied maps of the city until they could find home, school, grocery store, parks and restaurant. They used games, slides, and pictures to learn specific community objects such as traffic signs, food items, money, and directions. They went on two field trips (1) purchase items at the grocery store for a meal. (2) Eat lunch at Wendy's Restaurant. Since few students are able to communicate orally, they presented pictures to the clerk to identify the items they wanted. Mr. Alan photographed foods for the grocery store and restaurant, laminating them and giving the pictures to each students.

4

SKILL LEVEL

☐ Entry
☒ Adoption
☐ Adaptation
☐ Appropriation
☐ Invention

SAMPLE 7.4 The teacher prepares an outline of the project, its goals, and steps for completing the task and inserts them in the group portfolio.

tion to ensure the appropriate technology is available, and preparing an agenda. Figure 7.1 can be used as a planning guide for students and teachers to define the conference agenda. Figure 7.2 can be used to invite all interested parties to the conference, and Figure 7.3 can be used to assess the success of the conference. Electronic portfolios add flexibility to the conference. All parties can meet in the same location for the portfolio conference but do not have to meet face-to-face. Thanks to the technology, virtual conferences are now possible.

PLANNING GUIDE FOR PARENTS AND TEACHERS

1. Who will lead the conference?

2. Purpose for the conference:

3. What technology is needed to display the electronic portfolio?

4. Determine location and time:

5. Complete and send invitation:

6. Goals to be achieved during the conference:

    a._____

    b._____

    c._____

    d._____

7. How will it be determined if goals were met during conference?

8. Who is responsible for assessing conference outcomes?

9. What other materials, resources, or supplies need to be used during conference?

10. Who is responsible for collecting/setting up materials, resources, technology, and supplies for the conference?

Figure 7.1

CONFERENCE INVITATION

To: _____

Please plan to attend a portfolio conference for: _____

Date of Conference:_____ Location:_____ Time: _____

Purpose for this meeting is:

1._____
2._____
3._____

The conference will be led by: _____

Signature:_____ Date:_____

Please confirm attendance by returning this portion of the form by: _____
  ☐ Yes, I (we) plan to attend conference as scheduled.
  ☐ No, I (we) are unable to attend at the scheduled time; please reschedule.

Comments: _____
_____
_____

Signature:_____ Date:_____

Figure 7.2

PORTFOLIO CONFERENCE ASSESSMENT FORM

Student Name:_____ Conference Date: _____

Type of electronic portfolio used during conference: _____

Purpose of Conference:_____
_____

Attendees: _____

List goals discussed and outcomes determined:

1. _____
2. _____
3. _____
4. The most important outcome resulting from the conference was: _____
_____

Important Comments: _____
_____

Is a follow up conference needed: _____

Signature:_____ Date:_____

Figure 7.3

Plugging in the Portfolio

SKILL LEVEL

☐ Entry
☒ Adoption
☐ Adaptation
☐ Appropriation
☐ Invention

SAMPLE 7.5 Since visual pictures are an effective instructional strategy for many special needs students, the teacher guides the "buddy" and special needs student to obtain images from the Internet. Permission was granted from the businesses visited during the project. Images are saved on diskettes and imported into the electronic portfolio.

SAMPLE 7.6 This is a sample of how visual images can be used to increase understanding for special needs students.

Plugging in the Portfolio

SAMPLE 7.7 After at field trip to a local business, buddies and special needs students can document the event.

SKILL LEVEL

- [ ] Entry
- [x] Adoption
- [ ] Adaptation
- [ ] Appropriation
- [ ] Invention

SAMPLE 7.8 After surveying the students the results are placed in the electronic portfolio.

As electronic portfolios are increasingly used in conferences, they open doors for parents who would normally not visit the school for a variety of social, cultural, or economic reasons. Conferences can be held over the Internet, and communication can be conducted via e-mail.

Open communication and a well-planned agenda are the keys to successful conferences whether virtual or face-to-face. Students lead the conference, and they should begin the dialogue. They should first explain the purpose for using electronic portfolios and how the portfolios have changed the climate in their classroom. A short explanation is given to convey what is contained in the portfolio, why it was included, steps taken to complete the work, and how work was placed in the necessary format for the electronic portfolio. Students can begin navigating through the portfolio, demonstrating their technological expertise. Students can encourage parents and guests to ask questions or assume navigation of the portfolio. In this situation, the teacher acts as an observer, only participating if needed.

Plugging in the Portfolio

SKILL LEVEL

- [ ] Entry
- [x] Adoption
- [ ] Adaptation
- [ ] Appropriation
- [ ] Invention

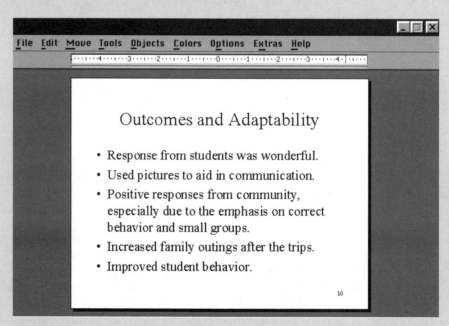

File   Edit   Move   Tools   Objects   Colors   Options   Extras   Help

Outcomes and Adaptability

- Response from students was wonderful.
- Used pictures to aid in communication.
- Positive responses from community, especially due to the emphasis on correct behavior and small groups.
- Increased family outings after the trips.
- Improved student behavior.

10

SAMPLE 7.9 Clear goals for the activity are placed in the electronic portfolio by the teacher.

Teachers and students should engage in open-ended conversations when they meet. Teachers ask prompting questions about why work was included, how the student felt about the assignment, how could it have been improved, and what tasks could be performed to deeper understanding. Listen to students' oral reflections about their work. Successful teacher/student conferences let students express their feelings both positive and negative. Allow students to show what they know!

When students are actively involved in the planning, discussion, and assessment of portfolio conferences, they transform classrooms from a teacher-directed to a student-centered environment. The more students participate in these processes, the more ownership they have in their learning. If private and public schools expect to produce workers with the skills needed for jobs in the information age, they must integrate technology across the curriculum, throughout instruction and assessment, where it becomes a seamless component.

The success of any electronic portfolio project is directly linked to a well thought out system for planning, organization, professional development, and strategies for implementation. Ongoing mentoring for both teachers and students gives the necessary support needed to ensure success. When electronic portfolios are embedded into the culture of the school, they make significant contributions to student achievement and teacher accountability by raising the expectations through "ownership" in learning for all students.

Equity and Diversity

Technology, like any rare commodity, is not equally distributed to every school, classroom, or student. It is increasingly important that students not only have access to a computer at school but that they engage in meaningful tasks when they are using it. Technology-enriched classrooms facilitate the following:

- Erasing geographic barriers
- Equalizing economic status
- Minimizing limitations of individuals

Not all students learn the same way; some readily process written information while others need visual images, repetition, or interactivity to gain an understanding of what is being taught. Because of the versatility technology brings to the classroom, teachers are able to customize curriculum, instruction, and assessment to create individualized learning environments. This opens doors for students, especially those with special needs to participate in projects such as electronic portfolios. Technology removes or minimizes barriers that prevent some learners from achieving their potential.

Students with Special Needs

The term "special needs" can be associated with "at risk" because traditional instructional strategies fail students who do not fit traditional concepts about learning. Adapting the learning environment within the regular classroom to include programs and practices for students with special needs more fully develops and helps mine one of the most valuable resources—young people. Adaptive technologies lessen the challenge of educating diverse learning disabled students and cultural or language minority students, as well as those considered gifted and talented.

Technology-appropriate learning tools compensate for verbal, visual, or tactile challenges students may encounter in the learning process. The sample portfolio shown in this chapter works with and relates to special needs students. Students with special needs are able to create electronic portfolios at their developmentally appropriate level without separating

Plugging in the Portfolio

SKILL LEVEL

☐ Entry
☐ Adoption
☐ Adaptation
☐ Appropriation
☒ Invention

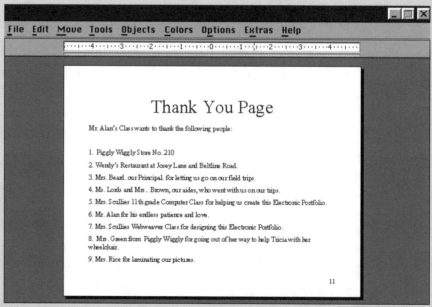

Thank You Page

Mr. Alan's Class wants to thank the following people:

1. Piggly Wiggly Store No. 210
2. Wendy's Restaurant at Josey Lane and Beltline Road.
3. Mrs. Beard, our Principal, for letting us go on our field trips.
4. Ms. Loris and Mrs.. Brown, our aides, who went with us on our trips.
5. Mrs. Scullies 11th grade Computer Class for helping us create this Electronic Portfolio.
6. Mr. Alan for his endless patience and love.
7. Mrs. Scullies Webweaver Class for designing this Electronic Portfolio.
8. Mrs. Green from Piggly Wiggly for going out of her way to help Tricia with her wheelchair.
9. Mrs. Rice for laminating our pictures.

11

SAMPLE 7.10 Students reflect upon their experience during a teacher-led discussion. The buddies and special needs students compile a thank you list for community and school members.

them from peers in the regular classroom. Computers compensate for challenges many students' experience while providing an infinite level of patience to repeat tasks until they are mastered or completed.

Students with Disabilities

Implementing a portfolio project with physically challenged students is manageable with technology. Students who are hearing-impaired can view the computer screen to design and lay out their portfolio. Menu items can be consistently placed on the desktop for easy navigation. Speech-impaired students can "talk" as voice-synthesizing software converts typed words into speech for auditory work. Students who cannot type information into their portfolio can speak into speech-to-text translation software. Braille keyboards and printers and image magnifiers enable visually impaired students to use technology, enhancing their ability to create an electronic portfolio. Alternative access methods allow physically impaired students to make selections by the touch of a screen or the use of specially designed keyboards or switches to participate in electronic portfolio projects.

Regardless of the disability, special needs students often feel inadequate and left out of many learning experiences. Electronic portfolio projects offer students a chance to be in control of their learning while gaining self-confidence and a positive self-esteem. Technology is allowing them to work alongside so-called "normal" students in an environment that is becoming increasingly user-friendly for all persons.

Students with Limited English Proficiency

The electronic portfolio is a valuable tool to demonstrate written and oral abilities regardless of a student's cultural or linguistic diversity. The visual feedback students receive from computers is especially helpful for students with limited English proficiency. Teachers can help students choose graphical software to help them reflect upon their work in ways that reflect their own culture.

Many word processing programs use a standard or special typeface, which allows students to express themselves in their native language and teach peers about it. Spell check, dictionaries, and thesauri are available in many languages. Recording oral presentations in native languages is beneficial in conferences to convey learning experiences to parents.

Gifted and Talented

Students described as gifted and talented may be at risk because many are not challenged by a traditionally paced curriculum. Engaging in an electronic portfolio project stimulates their creativity, enabling them to analyze information, formulate conclusions, and solve problems while assuming responsibility for the creation of the portfolio. Many intellectually gifted students have difficulty interacting with peers of average to low ability because the gifted students are perceived as uninteresting or lacking social skills. Engaging in collaborative electronic portfolio activities provides a positive social experience to develop interpersonal skills for all students.

Flexibility to Meet Student Needs

Electronic portfolio projects can be created at school or from remote locations. Students who cannot attend school because of extended illnesses and/or injuries benefit from the flexibility technology offers. Virtual interactions with the teacher and/or classmates can occur over the Internet. Peer and teacher reflections can be created when work is viewed online. Comments and replies can be transmitted via e-mail. Differentiating instruction is easily achieved for *all* students with the integration of technology.

APPENDIX A: UNIT PLAN

"The Oak Forest Nature Trail"

TOPICAL THEME

Preserving Our Environment—The Role of Nature in Literature

This school-wide project focuses on preserving the environment as it relates to character development and how things must change to adapt to different seasons and/or relationships. All classes within the K–6 school focused on different aspects of this yearlong project. All grade levels concentrated on experiences learned in literature and how they related to those in real life. Various writing activities throughout all grade levels focused on the use of comparison and contrast as a means of development in essay writing. A project template was developed by each team, noting the performance tasks in all disciplines and how technology is woven throughout as students use the knowledge gained in the project to build a nature trail on the school's campus.

Plugging in the Portfolio

The sample unit portfolio shown here shows how one second grade teacher implemented a whole class electronic portfolio project, creating an exemplary example for other classes to use as a resource. This class project focused on the concept of "change" through a study of "trees." Claris Homepage software was used to create the Web-based electronic portfolio.

SUBJECT/GRADE LEVEL

Reading/Literature K–6, Science K–6

CONCEPTUAL FOCUS

Character development/Change

PROCESSES

Language Arts, (Reading, Writing, Listening, Speaking) Creating, Computing, Thinking

MATERIALS

E. B. White, *Charlotte's Web*

Lynne Cherry, *The Great Kapok Tree*

Neighborhood MapMachine software by Tom Snyder Productions

The Graph Club software by Tom Snyder Productions.

INTRODUCTION

In some classes teachers read *Charlotte's Web and The Great Kapok Tree.* The class then discussed the environment and heroes in the literature. This served as a springboard for discussion of how things must change in order to survive and adapt to their environment. Any introductory discussions focused on what students should know and be able to do and understand upon the completion of the lesson and/or unit in order to extend their knowledge into their project work.

UNIT FOCUS

Discussions were directed toward "change" and how changes occur in the environment. Ms. Pennington's second grade class focused on a study of trees and how they change on the nature trail throughout the year.

TREE PROJECT CONTENT STANDARDS

English—Students should be able to speak and write appropriately for specific purposes and audiences.

English—Students should be able to access, organize, evaluate, and use information obtained by listening, reading, and viewing a variety of texts.

English—Students should be able to use literary knowledge as a basis for understanding themselves and society.

Science—Students should be able to use scientific process and inquiry methods (questioning, predicting, experimenting, collecting, and displaying information).

PROJECT ACTIVITIES

Students began their study by taking a walk on the nature trail and observing all of the different types of trees that live there. Leaves from every kind of tree found on the nature trail were collected and mounted for display in a class collection. Students generated, developed, and wrote comparison/contrast sentences/essays that dealt with issues raised in class discussions about how they thought the different trees began living on the nature trail. Students also researched other areas on the campus to compare/contrast trees in those locations to those on the nature trail.

The class made tree signs for each tree on the nature trail from materials donated by a local hardware store. These signs were one of several projects students made as part of this school-wide project.

ELECTRONIC PORTFOLIOS

The Oak Forest Webweavers created and maintained a school-wide electronic portfolio documenting how the Nature Trail was developed and what studies each class conducted. This multiage group of students met in the computer lab once a week to learn about Web authoring. They were also responsible for searching for educational Web sites related to specific topics that teachers needed to support their instruction. Webweavers were also asked by primary teachers to assist students with technology-related tasks.

Ms. Pennington's class decided to create a whole class electronic portfolio as part of the school-wide project. Linking it to the Webweaver's whole school portfolio, it became a resource for other classes to use when they studied trees. Mrs. Pennington arranged for the librarian to focus on research skills during their weekly library period. Students worked in teams during library time and project time to research different types of trees using reference books and the Internet.

Ms. Pennington led the whole class in the development of a storyboard for the construction of the portfolio. Each student team chose a type of leaf from the collection of tree mountings in the classroom. Working with the computer lab teacher, Ms. Pennington requested that students learn how to scan images using a flatbed scanner during their biweekly computer lab class. During the next several computer classes students learned how to use Claris Homepage in preparation to begin construction of their electronic portfolio.

Each team prepared a screen worksheet in class prior to going to the lab to use the computers. Any work that could not be finished during computer lab was completing during project time in the classroom. As the project progressed throughout the year, teams of students were asked to add information to the portfolio.

The culminating activities for the project was a presentation of a "Leaf Quilt" to the Principal at the Nature Trail dedication ceremony.

Integrated Activities from Other Classes

LITERATURE/MEDIA

ESSENTIAL UNDERSTANDINGS

All animals/insects work together to shape our environment.

All animals/insects must adapt to their environment in order to survive.

Experiences shared by characters in literature will parallel real life situations.

People may share similarities and differences with the heroes they admire.

GUIDING QUESTIONS

What is a true friend? Who is a true friend?

How do different animals/insects adapt to seasonal change?

How could the relationship be changed between Charlotte and Wilber?

What are the similarities between the main characters in *Charlotte's Web* and *The Great Kapok Tree?*

Plugging in the Portfolio

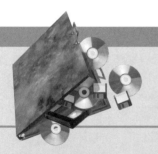

SKILL LEVEL

- [ ] Entry
- [ ] Adoption
- [ ] Adaptation
- [ ] Appropriation
- [x] Invention

SAMPLE A.1 This sample represents a whole class electronic portfolio showing a yearlong project. When placed on the district's Web site the portfolio can be used as a learning resource for others.

File Edit Move Tools Objects Colors Options Extras Help

Trees on the Oak Forest

Nature Trail

The "Penn-guin" second grade class at Oak Forest Elementary has been learning about the trees on our nature trail. We have identified some of the trees on our nature trail. Some of these trees may also grow in your neighborhood!

Scarlet Oak

CONTENT STANDARDS

Students should be able to speak and write appropriately for specific
purposes and audiences.

Students should be able to derive literal, implied, and personal meaning
from different kinds of texts and presentations (literary, informative,
and technical).

Students should be able to access, organize, evaluate, and use informa
tion obtained by listening, reading, and viewing a variety of texts.

Students should be able to use literary knowledge as a basis for
understanding themselves and society.

ACTIVITIES/TASKS

☐ Compare how and under what circumstances the animals communicated in *Charlotte's Web* with how they communicated in *The Great
Kapok Tree.* Discuss orally, make a Venn diagram, or write a compar-
ative essay or a sentence depending upon the grade level.

☐ Prepare, develop, and write a brief narrative essay recalling your
experience after taking a walk in nature. Younger children can
brainstorm as a group as the teacher writes a collaborative essay for
the whole class.

☐ Students sequence events in the story and create insect pictures
using a picture of themselves that displays a particular mood of a
character in the story. Students/teacher write a sentence describing
the mood. Students can research (Internet) insect information and
draw a picture of their insect using a drawing program.

☐ Working in cooperative groups using a laptop computer, write a
story about what the main character would do if he or she spent a
day on the school's nature area. One person in the group writes
about morning experiences, one about afternoon, and one about
evening. Another person writes the last paragraph, summarizing the
day. These activities can be teacher directed if younger students are
nonwriters.

☐ After reading Charlotte's Web or another piece of literature, prepare,
develop, and write an expository essay on the topic, "How would
our world be different if there were no animals/insects?" For
younger students who are nonwriters, this can be teacher directed.

☐ Select an insect such as a butterfly and research the different vari-
eties that exist. Create a newspaper section or butterfly brochure to
publish this information.

☐ Assign a character role to each group to develop and act out. Groups will interview each other's character and describe the character's importance in the story.

SOCIAL STUDIES

ESSENTIAL UNDERSTANDINGS

Life experiences and basic beliefs influence the actions of people and cultures in all environments.

Cultural differences and diversity can lead to peace/conflict influencing change in our environment.

Changes in population affect our environment.

Economic policies and growth affect the change in our environment.

GUIDING QUESTIONS

Why do people (or different countries) need friends?

How do life experiences influence the actions of people?

How do different cultures shape our environment?

How does our environment shape our life?

CONTENT STANDARDS

Students should be able to interpret social systems of different cultures, based on knowledge of their arts, religions, and philosophies.

Students should be able to analyze the impact of location and the interaction between the environment and people across continents.

Students should be able to make sound financial decisions, based on a knowledge of different economies and economic systems.

ACTIVITIES/TASKS

☐ Develop a timeline of significant events in the story line. Parallel these events with significant events in the real world that are familiar to students.

☐ Students draw a map illustrating a particular place described in the story. Determine where in the world the story is or could be set and research that area of the world, identifying climate, vegetation, and cultures.

☐ Students research and explore different cultures relating to the literature. Such a study should include customs, foods, geographical locations, and their arts.

☐ Divide the class into groups and have each group select a story element and make an oral presentation (videotape) to the class.

MATHEMATICS

ESSENTIAL UNDERSTANDINGS

Patterns of mathematics describe changes in the environment.

Methods of mathematical communication reflect and describe relation
ships, illustrate ideas, and apply principles of mathematics.

Mathematical reasoning shapes the environment.

Mathematical principles provide problem-solving strategies for real-world
issues.

GUIDING QUESTIONS

What mathematical patterns are evident in environmental change?

How do changes in the environment reflect friendships and behaviors
that exist in the real world?

Using mathematical reasoning, how can one solve problems, make
predictions, or create estimations in the environment?

How do methods of mathematical communication (models and
symbols) reflect changes in the environment?

Using principles of mathematics, how does one compute and use
different mathematical principles of geometry, measurement, and
algebraic function to solve real-world problems?

CONTENT STANDARDS

Students should be able to solve theoretical and real-world problems,
which require various approaches to investigate, understand, and apply
mathematical concepts.

Students should be able to express ideas and solutions through
appropriate mathematical language and symbols.

Students should be able to use mathematical reasoning to analyze
and answer theoretical and real-world questions.

Students should be able to make connections between mathematics and
their daily lives to answer questions, solve problems, and complete
authentic projects.

Students should be able to use appropriate technology to solve
problems and to communicate ideas and solutions.

ACTIVITIES/TASKS

☐ Students create a map of the barn and farm area in the story
Charlotte's Web. The map will be drawn to scale in the older grades
and include symbols with a color key.

☐ Students calculate the circumference of trees in the nature trail to
determine the age of the tree.

☐ Students take an observational walk on the school's nature trail. They collect different seeds for sorting, classifying, counting, and making patterns. Each group of students produces graphs displaying the different kinds of information they find. These graphs can be added to the class or school Web site.

☐ **Sorting**—Upon return to the classroom with the seeds collected, students are divided into groups of three to five. Students sort seeds by color, shape, and size. Each group creates a bar graph using graphing software to illustrate the results.

☐ **Classifying**—Upon return to the classroom with the seeds collected, students are divided into groups of three to five. Students classify seeds according to the plant the seed came from. Each group creates a bar or line graph using graphing software to illustrate the results.

☐ **Counting**—Upon return to the classroom with the seeds collected, students are divided into groups of three to five and count the seeds. They determine who had the most and least seeds within their group. Groups then compare the number of seeds among groups and produce a class graph. Each group then creates a bar or line graph using graphing software to illustrate the results.

☐ **Making Patterns**—Upon return to the classroom with the seeds collected, students are divided into groups of three to five. Students create various patterns with their seeds (for example, acorn seed, acorn seed, apple seed, and so on). Students then write a description of their pattern or explain the pattern to the teacher or peer.

SCIENCE/HEALTH

ESSENTIAL UNDERSTANDINGS

Changes in season affect the environment.

Cycles of nature influence changes in animal and plant patterns of growth.

Behaviors of diet and activity create a change in animal/plant growth.

GUIDING QUESTIONS

In what ways does seasonal change affect the environment?

How do cycles of nature influence changes in animal and plant survival?

How do plants and animals adapt to their environment?

In what ways has technology impacted the environment?

CONTENT STANDARDS

Students should be able to solve real-world problems through scientific
inquiry methods (questioning, predicting, experimenting, collecting
and displaying information, and drawing conclusions), using appropriate technology to communicate ideas effectively.

Students should be able to interpret situations that affect their everyday
lives by using knowledge of energy, matter, force, and motion.

Students should be able to use knowledge of the similarities, differences,
and interdependence of living things to analyze and assess events and
actions that impact life on Earth.

Students should be able to evaluate how science and technology affect
their personal lives and society as a whole.

ACTIVITIES/TASKS

☐ Students will work in cooperative groups of three to five to make
observations of insects, plants, and animals. They identify the external differences in their bodies making a Venn diagram to illustrate
their findings.

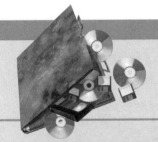

Plugging in the Portfolio

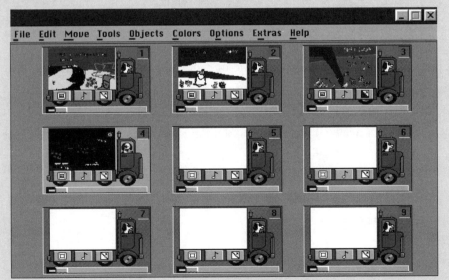

SKILL LEVEL

☐ **Entry**
☐ **Adoption**
☐ **Adaptation**
☐ **Appropriation**
☒ **Invention**

SAMPLE A.2 Using KidPix software to create a Slideshow Project Portfolio at the Primary Grade Level (K–2), students draw their picture to represent the season they are studying.

☐ Discuss climate and weather by looking at several different cities around the world. Choose several weather terms appropriate for the grade level. Have students discuss seasons and clothing changes for the different seasons. Discuss body coverings, size, and shape of a squirrel or other animal to determine how it survives the winter months. Discuss how it uses its covering to protect itself from the weather. Use drawing software or poster board to draw the animal on a nature trail during the winter.

☐ Have students identify animals and plants in a food web and describe their role (carnivores, herbivores, omnivores, insectivores, etc.).

☐ Demonstrate how several components in a food web can be affected when only one is disturbed.

☐ Draw a diagram identifying how the students' own lunch fits in a food web.

☐ Discuss a plant's role as a primary producer at the lowest level of a food chain.

☐ Take a trip to a nearby nature area. Observe, list, and compare components of the area's food web. If students do not see any animals, look for signs that they have been there—droppings, footprints, feathers, nibbled leaves and twigs, remnants of a meal (bones, fur, etc.), tunnel, or path. Discuss signs that people are part of a food chain. Fishing line caught in trees and empty shotgun shells are evidence of human predation. Does this area offer the natural resources needed for people to exist? What can be done to preserve the resources and food web?

☐ Have students draw a self-portrait at the top of a piece of paper. Below this, have them draw pictures of each item in their lunch and label them. Group each item in the different food groups and analyze if they are eating a balanced, nutritious diet.

☐ Students research deciduous and coniferous trees and annual, perennial, and wildflowers to determine what grows in the local area. Students calculate the amount of money needed to purchase tools, materials, and seeds for planting and then plant their seeds on the nature trail.

ART AND MUSIC

ESSENTIAL UNDERSTANDINGS

Works of art reflect cultural differences and diversity.

Composition (in art, music, literature) illustrates form and function shaping our environment.

Art communicates ideas, feelings, and attitudes about our environment.

Environment imitates art; art imitates the environment.

GUIDING QUESTION

How do art, music, and literature reflect cultural differences and diversity?

What is the relationship of composition to art, music, and literature?

How do the arts communicate environmental experiences (ideas, feelings, and attitudes)?

Does environment imitate art or does art imitate the environment?

CONTENT STANDARDS

Students should be able to communicate ideas, feelings, and attitudes through one or more of the art disciplines (dance, music, theater, and visual arts).

Students should be able to make sound judgments about the value of different types of artistic products (dance, music, theater, and visual arts).

ACTIVITIES/TASKS

☐ Students design a nature area for their school to improve the environment.

☐ Students brainstorm a list of winter clothing words. Distribute a collection of clothing picture pages among small groups. Have each child select one page and choose an article of clothing to cut out, then return to the large group. Demonstrate to the class a variety of ways to sort the pictures—color, size, and so forth. Encourage them to come up with other sorting rules—what one wears on their head, feet, body. Sort and glue pictures to a large chart. Take dictation or have students write using a word processing program to label types of clothing. Provide the prompt: Imagine you are on the nature trail on a cold winter day. Write and/or draw a picture to fit the phrase: "On a cold winter day, I took a walk and . . ." Have students use a drawing and word processing program to produce their work.

☐ After reading *The Great Kapok Tree*, discuss the names of the animals that spoke to the man. List them on the board or on butcher paper. Ask students to choose three animals from the list and determine if they could be found on the school's nature trail. Information can be gathered from the library, electronic encyclopedia, and Internet. Students can write a sentence or paragraph from the information they find and/or make an oral presentation to the class.

☐ Students create drawings depicting a significant event from one of the literature selections read and studied.

☐ Students analyze and critique selected songs, noting thematic and lyrical connections to literary selections read and studied.

☐ Students change and act out a different ending to one of the stories read and studied.

Plugging in the Portfolio

SKILL LEVEL

☐ Entry
☐ Adoption
☐ Adaptation
☐ Appropriation
☒ Invention

SAMPLE A.3 This illustrates how students displayed the information they collected. The images are scanned into an electronic format and placed in the Web-based application. Students work in teams to accomplish the task in class and computer lab.

LITERATURE FOR THE SCIENCE CURRICULUM (ELEMENTARY)

Birds

Ehlert, Lois. *Feathers for Lunch*

Heller, Ruth. *Chickens Aren't the Only Ones*

Hutchins, Pat. *Good Night Owl*

Oppenheim, Joanne. *Have You Seen the Birds?*

Pallotta, Jerry. *The Bird Alphabet Book*

Yolen, Jane. *Owl Moon*

Insects and Spiders

Brown, Ruth. *If at First You Do Not See*

Carle, Eric. *The Grouchy Ladybug*

 The Very Busy Spider

 The Very Hungry Caterpillar

 The Very Quiet Cricket

Dorros, Arthur. *Ant Cities*

Pallotta, Jerry. *The Icky Bug Alphabet Book*

Parker, Nancy Winslow. *Bugs*

White, E. B. *Charlotte's Web*

Plants

Bash, Barbara. *Desert Giant*

Behn, Harry. *Trees*

Carle, Eric. *The Tiny Seed*

Coucher, Helen. *Rain Forest*

Ehlert, Lois. *Planting a Rainbow*

 Growing Vegetable Soup

Guilberson, Brenda. *Cactus Hotel*

Heller, Ruth. *Plants That Never Bloom*

 The Reason for a Flower

Lobel, Anita. *Alison's Zinnia*

Merrill, Claire. *A Seed is a Promise*

Pallotta, Jerry. *The Flower Alphabet*

 The Victory Garden Alphabet Book

Silverstein, Shell. *The Giving Tree*

Ecology

Brown, Ruth. *The World that Jack Built*

Cherry, Lynne. *The Great Kapok Tree*

Cowcher, Helen. *Rain Forest*

Dorros, Arthur. *Rain Forest Secrets*

Gibbons, Gail. *Recycle!*

Greene, Carol. *The Old Lady Who Liked Cats*

Jeffers, Susan. *Brother Eagle, Sister Sky*

Oceans and Ocean Life

Baker, Jeannie. *Where the Forest Meets the Sea*

Cole, Joanna. *The Magic School Bus on the Ocean Floor*

Gibbons, Gail. *Whales*

Kalan, Robert. *Blue Sea*

Pallotta, Jerry. *The Ocean Alphabet Book*

    The Underwater Alphabet Book

Sheldon, Don. *Whale's Song*

Space

Asch, Frank. *Happy Birthday Moon*

    Moondance

    Mooncake

Barton, Byron. *I Want to be an Astronaut*

Plugging in the Portfolio

SKILL LEVEL

- [ ] **Entry**
- [ ] **Adoption**
- [ ] **Adaptation**
- [ ] **Appropriation**
- [x] **Invention**

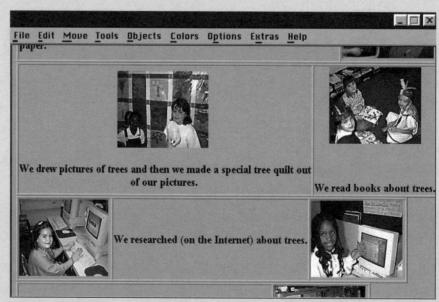

SAMPLE **A.4** Digital cameras can capture students working on their projects. It is important that safety precautions are used and district policy regarding Internet postings are followed (see chapter 3).

Brown, Margaret Wise. *Goodnight Moon*

Carle, Eric. *Papa, Please Get the Moon for Me*

Cole, Joanna. *The Magic School Bus: Lost in the Solar System*

Moche, Dinah. *What's up There?*

Simon, Seymour. *Jupiter*

  The Moon

  Saturn

  Stars

  The Sun

Internet

Animals

http://www.mindspring.com/~zoonet

Climate/Weather/Environment

http://www.weather.com

http://www.wetland.org/mshmarkt.htm

Insects/Spiders

http://www.scottyb.demon.co.uk/wildlife.htm

http://fiske.ci.lexington.ma.us/library2gr3kalambokas.html

Plants/Trees

http://cissus.mobot.org/MBGnet/vb/temp/index.htm

http://www.tcgcs.com/~nrolls/garden.html

http://aggie-horticulture.tamu.edu/wildseed/wildflowers.html

http://aggie-horticulture.tamu.edu/kinder/sgardens.html

APPENDIX B

Web Sites to Help Teachers Plug In

SITES OF HELP AND INTEREST

- Memphis City Schools' Teaching and Learning Academy Web site
  http://www.memphis-schools.k12.tn.us/admin/tlapages/on-line.htm
  Online tutorials, linked by subject matter. Offers computer tips.

- Media Builder
  http://www.mediabuilder.com
  Free image and font files, on-line font tools, and online imaging tools.

- San Diego City Schools
  http://projects.edtech.sandi.net/staffdev/tpfs98/index.html
  Staff development modules, online software tutorials.

- Macintosh Computer Tips
  http://users.desupernet.net/ohora/MacintoshTips.html
  Computer tips for Macintosh users.

- Windows 95:Beyond the Basics
  http://www.skwc.com/95/index.html
  Excellent and easy to understand desktop management training to use Windows 95.

INFORMATION ABOUT SOFTWARE

- Database
  Filemaker Pro
  http://www.edresources.com
  Microsoft Access
  http://www.microsoft.com

- Hypermedia
  Hyperstudio
  http://www.hyperstudio.com/index.html

Toolbook II
http://www.asymetrix.com

Digital Chisel
http://digitalchisel.com

Grady Profile
http://aurbach.com

Learner Profile
http://www.sunburst.com

Scholastic New Media
http://www.scholastic.com

- Multimedia

  Director
  http://www.macromedia.com

- Web Authoring

  Microsoft Frontpage
  http://www.microsoft.com

  Claris Homepage
  http://www.edresources.com

- PDF Documents

  Adobe Acrobat
  http://www.edresources.com

- Slide Shows

  KidPix
  http://new.broderbund.com

  Microsoft PowerPoint
  http://www.microsoft.com

- Office Bundles

  Microsoft Word
  http://www.microsoft.com

  Claris Works
  http://www.edresources.com

Technology for Teachers Sites

America Links Up

http://www.americalink.sup.org

Tips for Kids Online

National PTA Website

http://www.pta.org/programs/ftnight.htm

Right and Wrong Online – Teaching your Children Ethics in Cyberspace

U.S. Department of Labor

Secretary's Commission on Achieving Necessary Skills (SCANS)

200 Constitution Avenue, NW

Washington, D.C. 20210

http://www.workforceinvestmentact.com/scans/default.htm

GLOSSARY

3–inch diskette. Portable data storage device. 1.44 megabytes of information can be stored on one diskette.

Acrobat reader. Enables user to view and print PDF files from the Internet.

bitmap. File format for graphs. The file suffix is .bmp.

CD-ROM. Compact disc. A portable magnetic storage device.

cyberspace. Data created by the millions of online computers connected to the Internet. The term was coined by novelist William Gibson.

concept map software. A visual learning tool to help students prioritize, arrange ideas, and organize thinking through the use of outlines, graphic organizers, and webs.

database. Set of files containing related information. A logical set of files.

digital. Information stored in a numerical format.

download. To receive a file sent from another computer.

GIF. Pictures files stored with the extension of .gif. GIF stands for Graphics Interchange Format.

gigabyte. One thousand megabytes. The storage capacity of magnetic media such as floppy disks, hard drives, and memory.

hard drive. A magnetic storage device that may be part of a computer system, usually not portable.

HTML. Hypertext Markup Language. A set of ASCII characters that creates a hypertext document when embedded in a text document and interpreted by a Web browser.

hypermedia. Internet documents that consist primarily of hyperlinked sounds and images.

interactive. Two-way communication link between two or more computers in real-time.

Jaz drive. Large capacity removable or external magnetic storage device (larger than a Zip drive). Jaz is a registered trademark of Iomega corporation.

JPEG. Pictures files stored with the extension of .jpg. The name JPEG refers to the people that devised the standard: Joint Photographic Experts Group.

megabyte. One million bytes. The storage capacity of magnetic media such as floppy disks, hard drives, and memory.

modem. modulation/demodulation. A device that lets computers communicate over a telephone line.

multimedia. The capacity for presenting information in more than one format, including visual and auditory.

network. Two or more computers and/or peripherals connected by cables (wires) or wireless connections to allow communicating and/or data sharing.

optical recognition software (OCR). Enables text editing within a scanned image.

PDF. Portable Document Format used on the Internet.

platform (cross platform). Computer systems with unique architectures. At the classroom level, small personal systems are used. Two architectures exist: Macintosh and PC. When an application can be used on both architectures it is considered to be "cross platform" or hybrid.

RAM. Random-Access Memory. The part of a computer's memory used for documents and programs. RAM is erased when the computer is turned off.

server. A computer dedicated to servicing requests from users at a high rate of demand. Servers often have massive storage capacities and high processing speeds.

software. The programs and applications that run on the computer.

TCP/IP. The Internet is built on a collection of networks that cover the world. These networks contain many different types of computers. To ensure that different types of computers can work together, programmers write programs using a set of rules that describe how something is done (protocol). TCP/IP stands for Transmission Control Protocol and Internet Protocol.

TIFF. Picture files stored with the extension of .tif.

URL. Uniform Resource Locator. A naming or addressing protocol for computers connected to the Internet.

upload. To transmit a file on your computer to another computer.

WAV (Waveform). Format for sound files.

Web browser. Software that surfs the Internet. When the user provides a URL, a Web browser connects to a remote computer and displays its information.

Zip drive. Large capacity removable or external magnetic storage device. (smaller than a Jaz drive). Zip is a registered trademark of Iomega corporation.

REFERENCES

Ash, L., J. Luckey, and J. Avis. 1999. Outdoor achievement. *Design teams for school change.* Arlington Heights, IL: SkyLight Training and Publishing.

Ash, L. E. and J. Luckey. 1998. Outdoor achievement students build a nature center as part of an interdisciplinary curriculum. *The Science Teacher* April, 65(4):29–32.

Askew, J. 1998. Student portfolios and self-assessment rubrics found at *http://pc65.frontier.osrhe.edu/hs/science/ota4.htm*

Barrett, H. C. 1999. *Create your own electronic portfolio (using off-the-shelf software).* Found at *http://transition.alaska.edu/www/portfolios/toolsarticle.html*

Bellanca, J. and R. Fogarty. 1991. *Blueprints for thinking in the cooperative classroom.* Arlington Heights IL: SkyLight Publishing.

Branzburg, J. 1999. From paper to computers: The world of scanners. *Technology and Learning,* August, 20(1):41.

Burke, K., R. Fogarty, and S. Belgrad. 1994. *The portfolio connection,* Arlington Heights, IL: Skylight Publishing.

Burness, P., and W. Snider, eds. 1997. *Learn & live.* The George Lucas Educational Foundation.

CEO Forum Report on Education and Technology. 1999. *Professional development: A link to better learning year two report,* February, 1999.

Danielson, C., and L. Abrutyn. 1997. *An introduction to using portfolios in the classroom.* Alexandria, VA: Association For Supervision and Curriculum Development.

Erickson, H. Lynn. 1998. *Concept-based curriculum and instruction: Teaching beyond the facts.* Newbury Park, CA: Corwin Press, Inc.

Executive Office of the President. 1998. President's Information Technology Advisory Committee, *Interim Report to the President,* August.

Farr, R. and B. Tone. 1994. *Portfolio and performance assessment.* New York: Harcourt Brace.

Fogarty, R. 1997. *Brain-compatible classrooms.* Arlington Heights, IL: Skylight Training and Publishing.

Forcier, R. C. 1999. *The computer as an educational tool: productivity and problem solving.* 2d ed. New York: Prentice Hall.

Fountas, I. C. and G. S. Pinell. 1996. *Guided reading: Good first teaching for all children.* Portsmouth, NH: Heinemann.

Gardner, H. 1983. *Frames of mind: The theory of multiple intelligences.* New York: Basic Books.

———. 1993. *Multiple intelligences: The theory in practice.* New York: HarperCollins.

Gold, J. (Producer & Director) and M. Lanzoni. 1993. "Graduation by Portfolio - Central Park East Secondary School" [Videotape]. New York: Post Production, 29th Street Video, Inc.

Hibbard, K. M. et al. 1996. *A teacher's guide to performance-based learning and assessment.* Alexandria, VA: Association for Supervision and Curriculum Development.

Hughes O-Hora, C. J. *Macintosh tips and tutorials* available at *http://users.desupernet.net/ohora/index.html*

Lankes, A. M. D. 1995. *Electronic Portfolios: A New Idea In Assessment.* ERIC Digest ED39037795 *http//www.ed.gov/databases/ERIC Digests/ed390377.html*

National Center for Educational Statistics. 1996. *Pursuing excellence: Initial findings for the third international mathematics and science study.* Washington, DC: U.S. Department of Education.

Niguidula, D. 1997. *The digital portfolio: A richer picture of student performance. Studies on Exhibitions, Nos. 1-8.* Providence, RI: Coalition of Essential Schools, Brown University.

Purves, A. C., J. A. Quattrini, and C. I. Sullivan. 1995. *Creating the writing portfolio: A guide for students.* Lincolnwood, IL: NTC Publishing Group.

Ronis, D. 1999. *Brain-compatible assessment.* Arlington Heights, IL: SkyLight Training and Publishing.

SK Computing Solutions, *George and Mike's Guide to Windows 95* found at *http://www.skwc.com/95/index.html*

Sheingold, K. 1992. Presentation at a conference on Technology and School Reform, Dallas, June, 1992.

Simkins, M. 1999. Designing great rubrics. *Technology and Learning,* 20(1):23–24.

Slavin, R. 1995. *Cooperative learning: Theory, research and practice.* Needham Heights, MA: Allyn & Bacon.

Stigler, J. W. and J. Hiebert. 1999. *The teaching gap.* The Free Press: New York.

Teicher, J. 1999. An action plan for smart internet use. *Educational Leadership* 56(5):70–74.

U.S. Congress, Office of Technology Assessment. 1992. *Testing in American Schools: Asking the Right Question,* OTA-SET-519. Washington, DC: US Government Printing Office, February, 1992.

Viadero, D. 1994.Teaching to the test. *Education Week,* Extra Edition, July 14. Available at *http://www.edweek.com/ew/1994/41nsp.h13*

Wasley, P., R. L. Hampel, and R. W. Clark. 1997. *Kids and school reform.* San Fransisco, CA: Jossey-Bass.

Wong, H. K. and R. T. Wong. 1998. *The first days of school.* Mountain View, CA: Harry K. Wong Publications.

INDEX